BOUSSINGAULT

CHEMISTS AND CHEMISTRY

A series of books devoted to the examination of the history and development of chemistry from its early emergence as a separate discipline to the present day. The series will describe the personalities, processes, theoretical and technical advances which have shaped our current understanding of chemical science.

F. W. J. McCOSH

BOUSSINGAULT

Chemist and Agriculturist

D. REIDEL PUBLISHING COMPANY

A MEMBER OF THE KLUWER ACADEMIC PUBLISHERS GROUP

DORDRECHT / BOSTON / LANCASTER

Library of Congress Cataloging in Publication Data

McCosh, F. W. J. (Frederick William James), 1907–
Boussingault, chemist and agriculturist.

(Chemists and chemistry)
"Boussingault's scientific papers": p.
"Works of Boussingault published in book form": p.
Bibliography: p.
Includes indexes.
1. Boussingault, J. B. (Jean Baptiste), 1802–1887. 2. Agricultural chemists–France–Biography. I. Title. II. Series.
S584.52.B68M33 1984 630'.92'.4 [B] 84-4728
ISBN 90-277-1682-X

Published by D. Reidel Publishing Company,
P.O. Box 17, 3300 AA Dordrecht, Holland.

Sold and distributed in the U.S.A. and Canada
by Kluwer Academic Publishers,
190 Old Derby Street, Hingham, MA 02043, U.S.A.

In all other countries, sold and distributed
by Kluwer Academic Publishers Group,
P.O. Box 322, 3300 AH Dordrecht, Holland.

Printed in The Netherlands.

To Olga

. . . J'ai quatorze bouteilles
D'un vieux vin . . .
Bouçingo n'en a pas de pareilles.

[*Le repas ridicule* – Nicolas Boileau (1636–1711)]

'Quiconque professe l'économie rurale ou la
Chimie agricole a médité et médite encore Boussingault.'

[J. J. Th. Schloesing (1895)]

CONTENTS

LIST OF ILLUSTRATIONS

ACKNOWLEDGEMENTS

The gathering of material for this work has necessarily involved a search through a variety of sources in a number of institutions. My sincere gratitude goes to the staffs of libraries and archives for their willing and skilful cooperation. Such institutions are as follows:

Paris	Bibliothèque Nationale
	Muséum d'Histoire Naturelle
	Conservatoire des Arts et Métiers
	Académie des Sciences
	Assemblée Nationale
	Institut de France
	Hôtel de Ville de Paris
	Bibliothèque Historique de la Ville de Paris
	Académie d'Agriculture de France
	Conseil d'Etat
	Université de Paris à la Sorbonne
Saint-Etienne	Ecole Nationale Supérieure des Mines de Saint Etienne
	Société Amicale des Anciens Elèves
Unieux	Chateau Holtzer
Lyon	Bibliothèque Municipale de la Ville de Lyon
Besançon	Bibliothèque Municipale de Besançon
Strasbourg	Archives de Bas-Rhin
	Bibliothèque Nationale et Universitaire
London	British Library
	Royal Chemical Society
	Ministry of Agriculture
Johannesburg	University of the Witwatersrand
	Johannesburg Public Library
Bulawayo	National Free Library of Zimbabwe (for obtaining overseas works)
Harare	University of Zimbabwe (for general reference and use of microfilm reader)

Thanks are due to M. Charles Boussingault and Mme Françoise Boussingault, great-grandchildren of J. B. Boussingault for their hospitality – with Mme Charles Boussingault – and permission to cite extracts from letters and documents and for permission to publish photographs in their possession. Various institutions and publishers have kindly given me permission to use illustrations, photographs and quotations, e.g. Archives de l'Académie des Sciences de Paris, Masson, Conservatoire des Arts et Métiers, Longmans and Professor E. W. Russell for the frontispiece, John Murray, Muséum d'Histoire Naturelle, Imprimerie centrale administrative, Historiches Archiv Wetzlar, Attinger frères Neuchâtel, William Heinemann Ltd., which are gratefully acknowledged in the text. Permission to quote from previously published material has been kindly given by Taylor and Francis Ltd. (*Annals of Science*), Johnson Matthey and Dr L. B. Hunt (*Platinum Metals Review*) and Professor M. Gelfand (*Central African Journal of Medicine*). Photographs were supplied by M. and Mme G. Kapp of the château LeBel and such Alsace friendliness was extended by M. Fuchs who keeps the marriage and birth registers at Lampertsloch school.

Thanks are also due to Professor J. W. Perry, librarian of the of the University of the Witwatersrand for the loan of microfilm of Boussingault's scientific papers which were purchased at the suggestion of Dr W. C. Ahlers of the Dept. of Philosophy when I was living in Rhodesia. A grant from the Central Research Fund of the University of London made it possible to visit France for five months. Chapters of the Ph.D. (Lond.) thesis, from which this book evolved, were read and useful suggestions made by Dr F. L. Holmes (Univ. of W. Ontario), Professor R. S. Roberts (Univ. of Zimbabwe) and Dr W. A. Smeaton (Univ. Coll. London) to all of whom I am grateful, but especially to Dr Smeaton for his encouragement and advice over the past fifteen years. Although such generous assistance is acknowledged, I am entirely responsible for any error, mistranslation or misrepresentation of fact.

Westerham, Kent F. W. J. McC.

INTRODUCTION

It was the British music critic Neville Cardus, writing on Debussy, who remarked how "the great sea of Wagner threatened to overwhelm the world of nineteenth century music".[1] There is an analogy in mid-nineteenth-century agriculture where the great sea of Justus von Liebig developed a tidal wave which to this day conceals much of the original work and merit of others in the same field. Not only the general public but even students of agriculture may, or may not, recall the names of Persoz, Kuhlmann and Ville in France, Thaer and Sprengel in Germany, or even Lawes and Gilbert in England, to mention a few of them, whose pioneer works were not publicised in the same didactic and polemical manner as those of Liebig. Among such pioneers was Jean Baptiste Boussingault (1802–1887) whose fundamental researches contributed to the emergence of agriculture from an empirical corpus of facts to the status of a science. Yet apart from his work in animal and crop science he also engaged in metallurgical investigations, biology and pure chemistry. The scientific world was already approaching the end of an era in which it was possible to embrace several disciplines adequately. With increasing specialisation, institutionalism and professionalism in science the polymath was a gradually disappearing species.

Only once has a complete biography of Boussingault been attempted in which the author, a geologist,[2] succeeded in the space of fifty-two pages in mentioning the most important events of Boussingault's life and in describing his achievements, albeit briefly. This has proved a useful starting point in writing a more extensive biography.

His numerous activities have meant that a more satisfactory study could be made of any one facet of his life. The earliest of such memoirs were two objective accounts by contemporary agriculturists[3,4] but it was then too early to realise the basic significance of the investigations concerning nitrogen in the soil, plants and animals – what is now known as the nitrogen cycle in nature.

Although not a direct attempt at biography, mention is made of addresses given by Boussingault's colleagues at the Conservatoire des Arts et Métiers at the unveiling of a monument of Boussingault in the grounds of the Conservatoire.[5] Next appeared a chapter on Boussingault in a work about French agriculturists which is merely a *pastiche* of previous works except that mention is made of his first experience of the balance as a child when weighing out tobacco in his father's shop.[6]

The centenary in 1936 of the establishment of the first agricultural research station at his farm, Bechelbronn, in Alsace, was marked by the publication of a volume and several memoirs. The volume was a collection of Boussingault's contributions to discussions at meetings of the *Société Centrale d'Agriculture de France* which is a valuable record, best used as a reference book, of dated extracts in chronological order but divided into appropriate sections such as soils, crops, etc. The author recorded Boussingault's remarks, prefaced by a brief description of the subject under discussion, but without drawing any conclusions.[7,8]

At a centenary celebration at the Conservatoire, which followed a tour of houses in Paris occupied by Boussingault, an address was given by one of his successors.[9] Although an interesting synthesis it contained nothing new save a tribute to his teaching ability and to his gift for retaining the attention of his class, mainly by anecdotes of his South American experiences. Another centenary paper from *Izvestia* was notable for its comparison of the achievements and personalities of Liebig and Boussingault.[10]

For the following twenty-five years there appeared to be little or no interest in the work of Boussingault, partly accountable by the war of 1939–45. Among a series of lecture-discussions in 1961 at the *Académie d'Agriculture de France* was one on Boussingault and his colleagues which is noted for its freedom from biographical errors found in some previous papers.[11] An essay on his work in animal nutrition describes the various aspects of his work in this sphere but no attempt is made to relate it to a cycle of events centred on nitrogen and the rôle of forage crops.[12] This was the first paper where the text was related to a bibliography, but there was however too great a reliance on a secondary source which devotes a chapter to Boussingault's nutritional investigations.[13]

A recent contribution is a Yale doctoral thesis which firmly relates Boussingault's agricultural researches to the nitrogen cycle in nature, and is of inestimable value to a biographer. This is a most stimulating, authoritative and critical interpretation, and a reading of it suggests that a biologist rather than an agriculturist was required to assess the importance of Boussingault's agronomic work.[14,15,16]

In a volume on the rise of agricultural science in the U.S.A. there are chapters on the history of the subject which supply an excellent general background for anyone wishing to pursue the study further.[17]

It is hoped that this biography will lead to a greater understanding and appreciation of the multifarious interests of one whose aloofness fitted him for the life of a cloistered academic, yet experiencing more than the usual share of adventure.

Jean Baptiste Joseph Dieudonné Boussingault (1802–1887). (From *Soil Conditions and Plant Growth*, by Sir E. J. Russell (Longmans).)

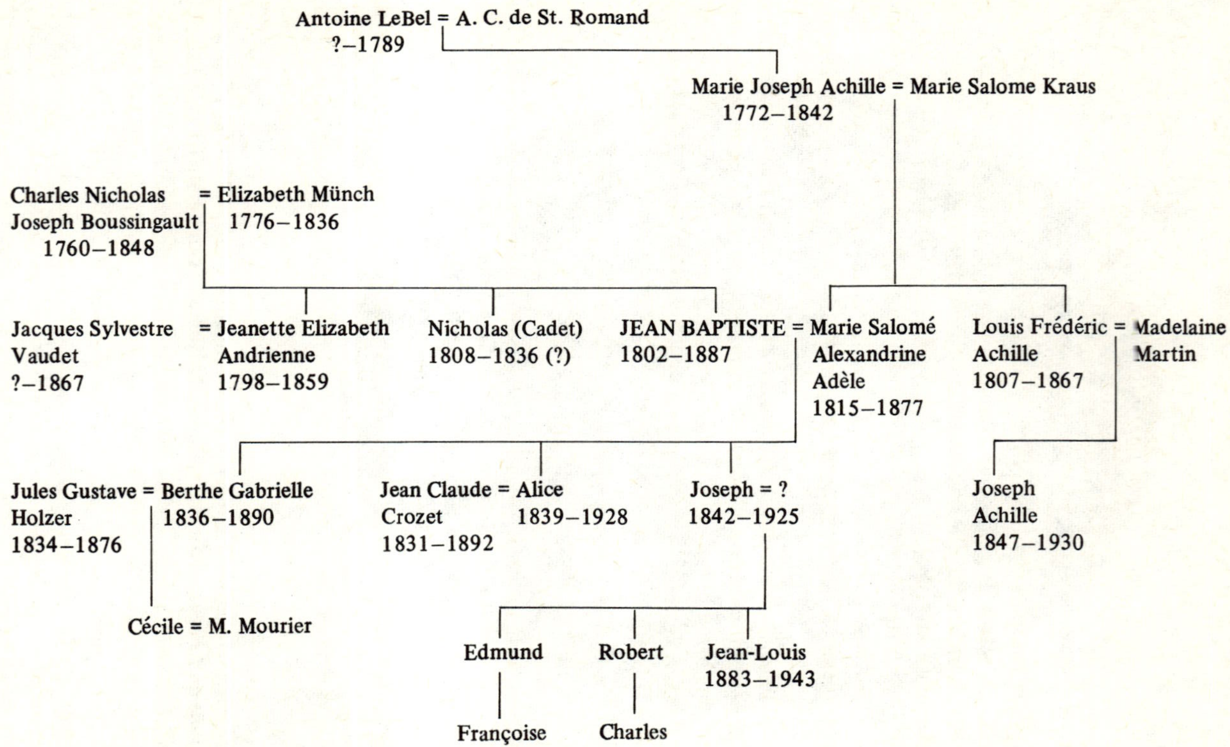

Fig. 1. The Boussingault and LeBel families – a partial genealogical table.

CHAPTER 1

THE BEGINNINGS [1]

Jean Baptiste Joseph Dieudonné Boussingault [2] was born in Paris on February 2nd, 1802. It was a brief time of transient peace, culminating in the Treaty of Amiens, which enabled Napoleon Bonaparte to establish his own political ascendency and that of the military power of France. To account for Boussingault's birth in a barrack room requires a modest familiarity with a Revolutionary war from which his father, Charles Nicholas, emerged as an army storekeeper. Charles Nicholas was one of a family of fourteen of whom eleven were girls.[3] His father held the prestigious title of Captain-General of the Royal Farms, an appointment where perquisites no doubt made up for a most unprestigious salary.[4] From the Captain-General onwards we can trace the strenuous attempts of the family to retain a foothold in the lower middle class of France, *la petite bourgeoisie*, to which end all fourteen children were educated at a friars' school in Orleans.

On the day when Charles Nicholas was reporting for his first duty at a Paris customs post in July 1789, the people of Paris, enraged by the imposition of internal duties on incoming goods, burnt down the customs posts.[5] His letter of appointment had been signed by the famous chemist Antoine-Laurent Lavoisier,[6] one of the Farmers General who purchased from the government the right of collecting the indirect taxes. Undeterred, he joined the Revolutionary army and was said to have risen to the rank of director of military hospitals of the army of Sambre-et-Meuse,[7] an improbable story for a man without medical training. More improbable when he is found in the fighting line at the battle of Wetzlar in October 1795 where he was slashed across the wrists by an enemy sabre.[8] Some senior rank, however, must have enabled him to recover from his wounds at the house of the burgomaster of Wetzlar, Andreas Münch,[9] There, at the Lutheran church on September 8th, 1796 'Carl Nicolaus Boisimcolt', according to the register, was married to Elizabeth, a daughter of Andreas.[10] From Wetzlar to Paris the journey was made by coach, ending in an anticlimax for Elizabeth Boussingault who had exchanged the rôle of burgomaster's daughter for that of a housewife to a barracks storeman in a city whose population spoke a strange tongue.[11] His demotion is unexplained – perhaps he overstayed his leave in Wetzlar? A daughter, Jeannette, was born in 1798, then came Jean Baptiste in 1802 and the youngest, Nicholas, was born in 1808.[12]

Fig. 2. Marriage entry of Boussingault's parents at the Lutheran church in Wetzlar, Sept. 8th, 1796.

Jean Baptiste Joseph Dieudonné – the choice of saintly names is not uncommon in France but it may have been influenced by an Article of the Concordat between Napoleon and the Vatican which stipulated that children were to be given saints' names only, or those of great personages of the past. The Concordat was an attempt to restore Napoleon's diminishing reputation after the disastrous capitulation of his army in Egypt in 1801. When a very small child Boussingault moved with his family to No. 20 rue de la Parcheminerie,[13] on the Left Bank, in a conglomeration of narrow streets which still retain a certain atmosphere redolent of the Paris of the early nineteenth century. It had, indeed, been a street of affluent parchment makers where sheepskins were prepared; but with the introduction of paper from China in the fifteenth century the street had gradually deteriorated.[14] It was a poor neighbourhood when the Boussingault family took up residence there, but there still remained some parchment makers among a group of colourful personalities which included an artist whose wife played the guitar in cafés, a colour engraver, a perruquier, wallpaper manufacturers and many other self-employed tradesmen and the inevitable bouquinistes, the booksellers of the Seine quays. Much of the immediate locality was changed beyond recognition when the boulevard Saint-Michel was pushed through these little streets in 1855, but the local playground, the churchyard of thirteenth century Saint-Severin, is still there. Father had now resigned from the army and was supporting his family with the proceeds from his tobacoonist's shop where Boussingault first learnt to weigh.[15] As shopkeepers the family was firmly entrenched among the *petite bourgeoisie* of the self-employed, which may account for Boussingault's belief in later years that an ancestor, Bouçingo, was the famous seventeenth century wine merchant of Nicolas Boileau's satire, *Le repas ridicule.*[16]

The friendly and extrovert life of the Paris streets did not then foreshadow the shy and introspective character of later years. With soldiers from the barracks he learnt to fence and collected stray kittens for the family physician, Dr J. LeGallois, whose studies of the spinal cord and its nerves were then becoming known.[17]

An Unconventional Education

With second-hand booksellers as his friends, especially M. Cordier who was godfather to sister Jeannette, there was no lack of interest in books when he was taught to read by a Mlle Lenormand who kept a dame-school.[18] Primary education took place at a school in the rue du Jardinet kept by a M. Deslyons where one of the pupils, Loubry, unwittingly started him on a scientific career several years later.[19] Boussingault then proceeded to the *lycée Louis-le-Grand* in the rue Saint-Jacques for a most undistinguished scholastic career:

> From the sixth class, where I understood little, I went to the fifth where I understood even less, then to the fourth and third where I understood nothing at all. The teachers treated me as if I were a blockhead. Nine-tenths of the pupils were in the same position. We passed from class to class like an iron bar in a rolling mill.[20]

He quitted school in the second class and therefore cast away any hope of entering a university. His was by no means an isolated case; there were many 'drop-outs' before reaching the *baccalaureat*. The commercial and industrial stagnation of France after the defeat of Napoleon in 1815 may have induced him to leave school and seek a job at once before the employment market was closed to him. It is a controversial point whether his lack of interest in school stemmed from the lack of science tuition. Napoleon has been quoted as one who supported the cause of science, and the service he rendered to the *Académie des Sciences* was undoubtedly greater than that given by Charles II to the Royal Society in the latter half of the seventeenth century, but such support did not extend to the lycées as only one-sixth of them provided science courses. Even this low figure is remarkable because Napoleonic educational thinking was against the teaching of science at lycée level, which accounts for the extremely elementary science courses provided by the universities.[21] Boussingault's inexplicable resentment towards school is exemplified by his reaction to father's request that he should write to the brewer and order beer. An impossible task, he considered, because at school he was compelled to read the letters of Cicero which were his only letter-writing models.[22] Although his class allegiance was now firmly fixed as the son of a shopkeeper and by attendance at a lycée it was, perhaps, a reaction to his sense of hopelessness that he was now playing in the streets around his home, and there he met his former class-mate of the primary school, Loubry. Loubry's mother was a washerwoman and among her clients were the professors of the *Collége de France*. One of them, Louis Jacques Thenard, employed young Loubry as a laboratory cleaner and so Boussingault agreed to help his friend,[23] but after a year he was discovered by Thenard who told him to go as he was too young for such work. Again, a year later, Thenard refused to admit him to his class of student-demonstrators, believing that his family was unable to support him throughout the course. Twenty-five years later, Boussingault became a member of the *Institut de France*, an honour corresponding in prestige to a Fellowship of the Royal Scoiety. "Ah!", said Thenard, "if only I could have foreseen".[24]

Unlike Loubry, Boussingault at once began to read physics and chemistry. His parents were wise enough to allow him to pursue his own interests and they abandoned their plans for making him either a pastrycook, then an affluent trade, or a pharmacist. Mother purchased for him the four volumes of Thenard's text-book of chemistry, *Traité de chimie élémentaire, théorique et pratique*, which appeared between 1813 and 1816, becoming the standard chemistry text-book in France for a quarter of a century. Those with scientific interests were extremely fortunate in the Paris of that time for there were excellent public lectures and the list of lecturers reads like a roll-call of the scientific genius of the early nineteenth century -- Thenard, Biot, Gay-Lussac, Haüy and Cuvier, among lesser known figures.[25] These were the Faculty of Science lectures primarily for those writing the examinations for degrees but they were attended by the sons of the aristocracy who thought that they had a duty to familiarise themselves with the progress of science. There were also foreign students but Boussingault belonged to none of these groups yet had the right to attend as did any member of the public. It is, therefore, not surprising that the lecture theatres were overcrowded and the students sometimes restive after standing for 2½ hours. Nor were the lecturers any less uncomfortable in a theatre where voices were raised because of the large concourse of sometimes over 500 students. Thenard also complained of the heat from his furnaces during the summer months, leaving, at the end of a lecture covered in perspiration.[26] Without any formal scientific preparation Boussingault attended lectures in chemistry, botany, crystallography, palaeontology and geology. "What a frightful hotch-potch of science", he afterwards observed, "after all I was only fourteen to sixteen years of age and there was no time to digest all that I had learnt".[27] Did this experience account for his polymath character in later life and the multiplicity of interests which he embraced? Geology appeared as a major interest encouraged by the purchase of mineral specimens from an old man, Soyer, known as "Prêchi-Prêcha" because of his incessant 'preaching' about the creation of the world. Soyer collected his specimens from those discarded from Haüy's laboratory and sold them from his stall on the parapet of the pont Saint-Michel, a kind of mineralogical bouquiniste.[28]

The period prior to Boussingault's free studies coincided with some of the most dramatic and disastrous events in French history. The family spoke in whispers on political topics for there was a division of opinion in Paris on the merits of Napoleon.[29] Men were recruited for the battle front never to return except those who survived minus a limb. So much was happening that he went out for the day with his lunch of bread and cheese, returning in the evening to tell the news to the family. Coffins of generals fallen in battle were carried along the near-by rue Saint-Jacques on their way to the Panthéon. The family was grieved by the death of Moreau at the battle of Dresden in 1813 as father had served under him; they resented Napoleon's accusation that Moreau had been an accomplice of Pichgru when the latter had defected in 1795. At the head of Boussingault's bed was father's sabre which he carried when under Moreau, engraved with Revolutionary ardour, '*Liberté, Egalité, Fraternité ou la mort*'.[30]

On the morning of 30th March, 1814 guns were heard as the allied armies of Russia, Prussia and Austria were advancing on Paris. Money and valuables were hidden in the cellar; Jeannette and some friends were lodged in a small room with mattresses placed at the windows. Paris capitulated the next day and Napoleon fled to Elba. The presence of foreign troops with money to spend gave a spurious air of well-being to the city. The masses held to the Republican ideal and were opposed to the Restoration of Louis XVIII as they feared a return to the domination of nobles and priests which was so characteristic of the *ancien régime*. Boussingault remarked on the arrogance of the clergy, and the 'devout' in the neighbourhood regarded his Protestant mother as a heretic. For the sake of appearance he made his first confession at Notre Dame although he felt neither Catholic nor Protestant. The return of Napoleon from Elba, and his defeat at Waterloo, meant that an army, mainly Prussians, required billeting in houses, when Boussingault's mother was busy as an interpreter, going from house to house in their quarter – possibly the most sociable period of her married life.[31]

The reconciliation between the supporters of the Bourbons and the Bonapartes, which lasted during the Hundred Days, was broken at the second Restoration of Louis XVIII, and those who had sided with Napoleon were hunted and executed, the most famous of them being Marshal Ney. Boussingault had a good view of some of these executions from a room in the Palais de Justice where lived the Benoist family, for M. Benoist was director of the archives there, and his son was a friend of Boussingault.[32]

He Enters a Mining School at Saint-Etienne

When Boussingault was sixteen a decision on his future career was imperative.

He was now fired with the zeal of overseas travel and first enquired about prospects in the French navy, but as aristocratic birth was a prerequisite for promotion he turned his attention to the Russian navy several of whose warships were now captained by Frenchmen. Alas, the language proved too difficult.[33] Lack of aptitude in learning a foreign language appears remarkable in a bilingual family yet difficulty there was throughout his life. When in Alsace we find him imploring his German-speaking mother to keep house for him because he could not understand the locals.[34] His attempts to lecture in Spanish in Colombia may have induced the government to transfer him to mine inspection and prospecting.[35] Much later he is declining a request to visit Rothamsted in Hertfordshire, Mesdames Lawes and Gilbert offering their services as interpreters.[36] In spite of his adventures during the rumbustious days of Simon Bolivar he was basically a shy man and his introvertness may have contributed to his incapacity to learn a foreign language.

At this time he heard of a practical school for miners which had just opened at Saint-Etienne near Lyon. A certain degree of optimism was required of applicants to mining schools for the history of such institutions in France was most discouraging. Two such schools were to be established about the middle of the eighteenth century, one at Forez for coalmining and the other at St. Marie-aux-Mines for the mining of metals, but a lack of applicants caused the proposals to be scrapped. Applicants wished to be trained as artisans but it was the policy to regard technical courses as suitable for handymen. The *Ecole des Mines* was founded in Paris in 1783 but the Minister of the Interior decided that the courses were insufficiently practical and closed the school in 1802 only to reopen it in 1816 for those who were of the standard of a graduate of the *Ecole polytechnique*, which may have decided Boussingault to apply for the more modest school at Saint-Etienne. In place of the *Ecole des Mines* closed in 1802 two mining schools were opened in the same year, at Kaiserslautern in the Sarre and at Pessy in Savoie, both then within the the borders of France. By the treaty of Vienna of 1815, following the defeat of Napoleon, the Sarre and Savoie were no longer a part of France and so a new mining school was started at Saint-Etienne in a house in the rue de Roanne to the north of the town. It moved in 1843 to the château de Chantegrillet and was transferred to its present site in the cours Fauriel in 1928. The apparatus, library and mineralogical collection came from the school at Pessy, and the director of the school, M. Beaunier, had been head of the school at Kaiserslautern.[37] Whether Boussingault had ever intended becoming a miner is extremely doubtful but he was attracted by the prospect of working in a laboratory, making use of a library and examining the mineralogical collection. At the same time he persuaded his friend Benoist to apply.[38]

From Paris, in November 1818, they walked towards the Saône through Lieusaint, Montreau, Sens, Joigny and Auxerre, meeting the river at Chalons-sur-Saône which they followed to Lyon, and thence overland to Saint-Etienne, arriving during the first week in December.[39] They travelled light and a trunk sent on in advance contained the following:

1 frock coat	1 neckcloth, cotton cambric
1 pair trousers	4 handkerchiefs
1 waistcoat	4 pairs cotton stockings
3 shirts	1 fancy waistcoat
4 neckcloths pink	2 pairs shoes
1 neckcloth lilac	1 night cap
1 neckcloth, white embroidered	

All this is known because the trunk was lost en route and a list of contents was sent by father.[40]

The staff of the mining school were intensely practical men with interests in local industry. The director, Beaunier, a metallurgist, had established the nearby steelworks of La Barardière, and later directed the construction of the first French railway, bringing coal and iron ore from the Haut-Loire basin to Saint-Etienne.[41] Other staff were Gallois (ferrous minerals), Gueynivaud (chemistry), Desroches (metallurgy), Thibaud (mining engineering) and Burdin (engineering drawing).[42] Classes were limited to eight or nine students in each of the two years of the course. In the inaugural class of 1817, before Boussingault and Benoist arrived, was Benoît Fourneyron who invented the water turbine in 1834 and was afterwards related to the Boussingaults through marriage of his nephew, Jean Crozet, to Boussingault's daughter Alice.[43] The cost of the mining course is not known except that tuition was free, but his allowances and expenditure are revealed in letters expressing financial uneasiness and anxiety to which he attributed his insomnia. His self-confidence was at low ebb when he suggested withdrawing from the school to enter a commercial house or factory. Father promptly countered this proposal by threatening to remove him if his health was really affected, and, no doubt, thoughts of the laboratory, library and mineralogical collection quickly dispelled such defeatist ideas.[44] Henceforth, his letters were more cheerful but too objective for his parents, filled as they were with details of his school life when they wished to read about Lolo's[45] mode of living, as he was known in the family. Economies he did make for he writes,

As for drink, there is a fountain at our door and I can assure you that I have never drunk so much water in a town which promised wine in plenty.[46]

The curriculum included geology, metallurgy, mechanics and practical mining in the first year. In the second year the subjects were statistics, geometry, mechanics applied to mining operations, chemistry, mineralogy, machine drawing and further practical mining. The weary, uncomfortable and sometimes dangerous conditions in the mine did not discourage him:

Alas, if those Parisians could see us leaving the mine, pick and lamp in hand . . . dirtier than the blackest chimney sweep. If they penetrated the underground workings to great depths so that they could see us lying sometimes on our stomachs, or on our backs in water and sometimes mud . . . we laugh at ourselves.[47]

There is little mention of textbooks. The second edition of Thenard's *Traité de chimie* . . . was probably a personal choice, but he ordered that best-seller among mathematical texts, Legendre's *Eléments de Géométrie* published in 1794 of which the thirteenth edition appeared in 1871. Altogether he appears to have settled down to a satisfactory student life where the curriculum did not extend him either physically or mentally.

Vacation Assignments

The Lyonnais are extremely fortunate in having in their immediate vicinity such magnificent and mountainous scenery to which, today, they take themselves by car on Sundays. To the north-west are the monts du Beaujolais, dappled yellow with broom in the month of May: to the west are the monts du Lyonnais which end to the north of Lyon with the most d'Or. Of interest to the historian of science is the homestead museum of the physiologist, Claude Bernard (1813–1878), at Saint-Julien in the Beaujolais, and another such museum of the physicist, André Marie Ampère (1775–1836), at Polymieux near the mont d'Or. The road from Lyon to the Beaujolais passes through Chessy-les-Mines, about twenty-two kilometres from Lyon, and it was to Chessy and nearby Saint-Bel that Boussingault and a fellow student were sent in the Easter holiday of 1819 on a project assigned by the staff of the school. Up to the middle of the nineteenth century the copper mines at these two places yielded a beautifully coloured blue ore known as Chessylite or Azurite, a basic copper carbonate. The assignment was probably given by Thibaud, the lecturer in mining engineering, who subsequently published an article on the subject of the Chessy ores.[48] From the two students answers were required to a questionnaire which suggests that the Saint-Etienne teaching embodied an approach as modern as that found today. The ecological and sociological aspects of the mining industry were considered in such questions as,

What trade outlets are there for the sale of copper?

What political policies are favourable or unfavourable to this sale?

How does copper mining affect agriculture and commerce?

What is the influence of copper mining on the morals of the inhabitants?

The students were not taking any risks over the last question but they consulted the local curé who "gave all possible help, as much as could be suitably allowed, on the morality of both sexes", a strange case, perhaps, of the confessional aiding the cause of scientific education.[49]

In August, at the end of the school year, he returned to Paris with friend Benoist, not by the direct route but via the Auvergne with its volcanic rocks. First, however, they examined a local phenomenon, the burning coal-mine at Ricamarie, between Saint-Etienne and Firminy. The fumes contained the sublimation products of sal-ammoniac which was deposited on the wall of a peasant's cottage centrally heated by the burning mine. The dry distillation of coal produced ammonia but what was the source of the hydrogen chloride gas which was necessary to form sal-ammoniac? This query formed the subject of a letter which was subsequently published in *Annales de chimie et de physique*, Boussingault's first contribution to a scientific journal although he modestly attributed it to "the student miners of Saint-Etienne".[50] From Ricamarie they proceeded to the extinct volcanoes of the Auvergne, climbed the Puy-de-Dôme and recollected that Florin Perrier had climbed it in 1647 carrying a mercury barometer to show that air pressure falls with increasing altitude. Their protracted return to Paris had left them with little money – "We sheltered in impossible inns; our finances were very low; the geology of the Auvergne had ruined us".[51]

The Parisian environment was a disillusionment after the mountains of the Forez and the Auvergne; the restricted life of the capital compared unfavourably to the wide horizons of the Midi. The return journey from Paris to Saint-Etienne in October 1819 followed the route taken in the previous December but a detour was made through Autun to examine the coal mines at Le Creusot where their former mining school colleague, Benoît Fourneyron, had been appointed on leaving Saint-Etienne.

He Becomes a Student-Demonstrator

The holiday investigations at the copper mines of Chessy and Saint-Bel have been mentioned. Boussingault was recalled from these assignments to prepare lessons for the head of the laboratory, M. Gueynivaud; presumably this meant the material and the apparatus. This was the starting point of a series of embarrassing events. There was some incompatibility between Thibaud and Desroches, who taught metallurgy, as a result of which Thibaud was dismissed, a decision which did not please Boussingault who had collaborated with Thibaud in experiments on the fermentation of potato starch during the winter of 1819–1820. Beaunier, the director, replaced Thibaud by Boussingault who was given the title of student-demonstrator or *élève-breveté*. Boussingault was furious and threatened to follow Thibaud to Grenoble where he had obtained employment, but Beaunier dissuaded him by suggesting attractive projects in the laboratory.[52] Further, the status of student demonstrator gave him staff privileges which were resented by some of the student body.[53] One such privilege was the free use of the laboratory where he was to be found on most days, even on Sundays, and together with other members of staff, all experienced in the chemical industry, he pursued several projects which were more relevant to chemical engineering that mining or geology. He had already cooperated with Thibaud in fermentation experiments which were now extended to the production of glucose from grapes, cane sugar, and even from wool, flax seeds and rags because the distillers in Lyon were seeking cheap sources of glucose for the hospitals. He even offered to direct the glucose factory and to establish a metallurgical shop for the production of superior files, to be supervised by his brother-in-law, Sylvestre Vaudet, who had married sister Jeannette.[54] His attempts at entrepreneurship were encouraged by a naive belief that local industry would immediately sponsor factory scale reactions which had yet to be tried beyond the laboratory stage.

The Rôle of Silicon in Steel: A Silicide of Platinum is Discovered

M. Beaunier was director of a steel foundry near Saint-Etienne and in steel samples from the foundry Boussingault had demonstrated significant percentages of silicon, from which he suggested a theory, since found untenable, that the properties of steel were due, not to the presence of carbon as was generally believed, but to the silicon content. Ancillary to this research was an attempt to produce a platinum-carbon alloy, or a platinum carbide analogous to steel, by heating thin leaves of platinum with wood charcoal in a clay crucible. The occasion was almost catastrophic as can be gathered from a letter to father;[33]

Since my last letter I have seen my cousin, Fouché-Séguimard, I have set fire to the school and I have melted a metal (platinum) which was believed to be infusible.

My cousin came to see me the very day my furnace was lit; he spent several moments with me in the laboratory with a local gunsmith and M. Beaunier who had come to see me at work. The heat of the furnace was extraordinary and, according to M. Beaunier, was greater than that found at his steelworks. After three hours firing I drew the fires from the furnace, closed all the dampers of the flue and left. The next day, impatient to know the result of my experiments, I was in the laboratory at five o'clock. I met a teacher who said he had smelt an odour of burning throughout the night. We entered the lecture room and hardly had we opened the door when flames were seen in the midst of the smoke which filled the room. I at once called the caretaker, gathered together the student miners and in two hours we had mastered the fire, and the insurance company paid for the damage.

The cause of the fire was that the flue had not been constructed for a fusion furnace, and a beam which supported the floor of the first floor, being too near this weak flue, caught fire in the night. Fortunately the door of the lecture room was closed, and more fortunately I had taken the trouble to close the furnace before leaving, otherwise it is certain that the mining school would no longer exist.

As soon as the fire was extinguished I went to look for my crucibles, and had the satisfaction of seeing that my platinum had melted; in another crucible I found that the platinum had combined with carbon and formed a casting similar to an iron casting.

His enthusiasm had bounded ahead of the facts. First, platinum had been melted as far back as 1758 by Macquer and Baumé; also, he was to find subsequently that he had prepared, not a platinum-carbon alloy, but a silicide of platinum which was more fusible than the metal. Erroneously he asoribed the formation of a silicide to the presence of silica ash in the wood charcoal. Over fifty years later he was to find that the silica came from the walls of the crucible. It was unusual, and still is, for the work of a nineteen year old student to be accepted by a foremost scientific journal but the *Annales de chimie et de physique* published his findings in 1822.[56]

Any confusion as to whether the properties of steel could be ascribed to carbon or silicon was due to two factors; first, it was believed that they were due to the presence of carbon; secondly, the production of carbonless steels suggested that another element was responsible. Louis Clouet (1751–1801), a metallurgist, had produced a steel in 1798 with apparently no carbon but he was too hesitant to suggest that it was possible to recognise a steel without carbon.[57] Boussingault carefully followed Clouet's instructions and produced a steel with a small percentage of silicon but with no carbon. Beaunier, on a visit to London in 1821, communicated Boussingault's results to William Wollaston (1766–1828) the English physician whose gravimetric and volumetric investigations of salts pointed to practical proof of the existence of atoms which had hitherto been a theoretical concept from the Greek Democritus down to his contemporary John Dalton. Wollaston was interested in these results but correctly suggested that insufficient care had been taken to test for carbon in the Clouet steel.[58] Beaunier also showed the results to Faraday and his cutler assistant, Stodart, but both thought that it was impossible to replace carbon in a steel by silicon even though Stodart had prepared carbon free steels using iridium or osmium.[59] The solution to this frustrating problem was simple. When Boussingault returned to these experiments many years later he had the advantage of more advanced techniques and more critical methods of analysis, which proved that the so-called carbonless steels – even Clouet's steel – contained a small percentage of carbon derived from the furnace gases during their preparation, enough to give them the malleability of a mild steel. Whatever initial errors there may have been in the interpretation of his experimental results, it was a remarkable piece of research by an eighteen year old whose formal training in chemistry had started eighteen months previously, not in a university but in a mining school. Yet in spite of such an auspicious start it is a matter for conjecture whether the chemistry course at Saint-Etienne was sufficiently systematic – or even systematic at all. Without a knowledge of the school syllabus no clear answer can be given to this question. His attendance at Gay-Lussac's lectures, prior to entering the mining school, may have implanted the concept of chemistry as a subject of general laws but if, as he admitted, the free lectures in Paris left him confused, he may not have appreciated the need to seek general laws concerning the relations between chemical phenomena, or to care very much about them. A few years later, in South America, he is claiming the presence of 'compounds' in alloys of gold and silver found in ores. This, and other examples, more than hints that, in spite of a long life of excellent and meticulous analyses, a plausible reason for his failure to become a truly great chemist was the lack of a systematic approach in his chemical studies at Saint-Etienne. Perhaps this can be simply stated by saying that he suffered from the lack of a university training.

His Character Development at Saint-Etienne

During these two years there developed in Boussingault certain character traits and attitudes which were to manifest themselves throughout his life. They may have been evident before he left home but in the absence of any independent commentary on his life in Paris up to 1818 we must rely on his correspondence from Saint-Etienne. Politically, he was influenced by the events of his time. The liberal Constitutional Charter, introduced by Louis XVIII on his Restoration in 1814, had almost disappeared as a result of reactionary legislation. The assassination of the heir to the throne, the duc de Berry, in February 1820, provoked press censorship which in turn led to pro-Charter disturbances by students which were quelled by the Royal Guard with loss of life. Boussingault endeavoured to sustain the imaginary and unrealistic rôle of political detachment expected of scientists, but rarely achieved, yet at the same time displaying the reactions of his own *petite bourgeoisie* in blaming parliament, individual liberty and the freedom of the press for the national malaise.[60] Business difficulties came in the wake of the assassination and Paris languished in a trade recession. The tobacconist shop in the rue de la Parcheminerie now closed early.

In his letters from Saint-Etienne, Boussingault exhibited a self-centred personality, rarely mentioning his fellow students. A somewhat withdrawn attitude, which was to appear often in his life-time, is noted in his correspondence which included criticism of the lively behaviour of other students, and he was sufficiently formal to write:

> One easily distinguishes two classes of students at our school because upbringing is not the same for all. Each one mixes with those who suit him best. As for me I have attached myself to the refined students which I have already mentioned. Benoist mixes with the others and has kept up some of the Paris customs which are unsuitable here.[61]

M. Benoist at the Palais de Justice was saddened by the break between Boussingault and his son and expressed his concern to M. Boussingault.[62] It appears doubtful that the type of upbringing of his fellows was known and he does not clarify his remark about the customs of Paris. Which of the refined students were his associates is not known, for apart from professional friendships with staff members he does not appear to have formed any attachment. Perhaps his dual rôle of student and teacher encouraged this estrangement from the main student body? From April 1820 his special status may have aroused jealousy for he refers to his being shielded from intrigue and calumny, again without details.[63]

His correspondence to his family became more and more infrequent, requiring a gentle reprimand from father especially when, on one occasion, he failed to acknowledge the receipt of his monthly allowance.[64] The voice of protest, however, belonged to his sister, Mme Vaudet, and his brother Nicholas who required personal replies to their letters and were not content to glean news from letters addressed to the parents. Mme Vaudet, his sister Jeannette, wrote:

I want at least as much as you do, to have a regular correspondence. I am sorry that it starts at a time when it becomes difficult and almost impossible. This is the second letter you have written to me in three years; you should not talk of other people's laziness.[65]

For the Boussingault parents, religion had always been a matter of expediency rather than loyalty to a faith, for this was a satisfactory compromise when father was nominally Catholic and mother a Protestant, but it led to extraordinary results in the children, namely two ardent Catholics and one agnostic. The sister, Jeannette, became a fervent Catholic and from an early age she attended Low Mass at Saint-Séverin each Sunday, accompanied by a none-too-willing Boussingault.[66] The younger brother, Nicholas, who was physically weak and perhaps a little sub-normal mentally, was also strongly Catholic and conveyed by letters to his brother at Saint-Etienne the family resentment towards his uncomplimentary views on Catholics. On one occasion, having attended a service at nearby Saint-Chamond, Boussingault wrote,

What do people think of the priests in Paris? Here they are the laughing stock of sensible people. I have heard them preach at Saint-Chamond. I shall talk to Vaudet about these prayer sellers.[67]

A domestic crisis occurred when he was asked to attend the baptism of his sister's child and become a godfather.[68] He refused as this would have meant an absence of several weeks from the mining school towards the end of the final year. It was, of course, a Vaudet child and the angry reaction of the Vaudets was conveyed in one of the rare letters from mother who was not upset but merely asked her son to let her know when he would be next in Paris as she would be very pleased to see him.[69] Boussingault's cynicism and sarcasm concerning religion were not merely the voice of youthful protest for he had grown up during a period in which the forces of reaction were supported by the Church and therefore his criticism of insincere religious practices was not unexpected.

Saint-Etienne was a period of scientific and personal development for him.

Intended by training for a career in coal-mining, the scientific and especially the chemical aspect of the mining school revealed to him a path which he eventually pursued, that of the scientist rather than that of the mine manager. Although on leaving the mining school more attention was paid to social interests and his domestic life, yet the initial financial and social anxieties of the early Saint-Etienne days, coupled with an asocial tendency, developed traits which were to become reinforced in his later years. Altogether, Saint-Etienne must have proved more than just a satisfactory phase for his director, M. Beaunier, to have written:

You are among the small number of my disciples who have given more to the school than you have taken out of it. Have a thought sometimes for an institution where you have left behind so many happy memories.[70]

CHAPTER 2

ALSACE AND THE NEW WORLD

Towards the north of Alsace is a neat little town, Merckwiller-Pechelbronn, situated on the Pechelbronn oilfield. Travelling northwards from Merckwiller, as the town is known locally, the road climbs, crosses the disused rail track of the former oil refinery, then passes through the pleasant and small suburb of Cité LeBel which was formerly the residential area for professional and administrative staff employed at the refinery. The road then descends to the attractive village of Lobsann with its half-timbered houses, most of them gay with window-boxes. It was at a lignite mine, two kilometres from Lobsann, that Boussingault found his first employment. Although at the time he could not have realised it, the acceptance of this post meant far-reaching consequences, both professional and personal.

There was no lack of employment for those leaving the mining school in 1820 but first he was tempted to join Thibaud at Grenoble until the mining school director, Beaunier, induced him to accept a mine manager's post at Lobsann. Apart from lignite, the mine yielded bituminous sand from which petroleum could be distilled. Perrier et Cie were the owners and they proved willing to pay Boussingault's expenses on a tour of France to inspect mines and factories. It appears that the production of sulphuric acid was contemplated, perhaps from pyritic ore, and the possible manufacture of alum and iron sulphate or copperas.[1]

During this tour he was bold enough to visit the illustrious scientist Joseph Louis Gay-Lussac (1778–1850)[2] who lived at the Arsenal in Paris where he was chemist to the *Commission des poudres et salpêtres*, an important body concerned with the production of gunpowder. He is known mainly for his work on the combination of gases by volume, and for a vast amount of original work in the field of analytical chemistry. His early life was an example of successive triumphs over seemingly unresolvable adversities. The Revolution broke out when he was a student in Paris, his course being closed and the students sent home with the exception of Gay-Lussac who was asked to continue because of his outstanding mathematical talent. His father had lost his position as a magistrate and could not support his son at university, but by giving private lessons he continued his studies and was made an assistant to Claude Louis Berthollet (1748–1822) a pioneer of physical chemistry. In spite of his shyness, his introspection and the resulting solitary life, Gay-Lussac was adventurous as can be concluded from his two balloon ascents to 4000 and 7000 metres in 1804 in order to demonstrate that there was little change in the composition of the atmosphers as one ascended.[3] From Boussingault, Gay-Lussac received a copy of his memoir on platinum silicide which he promised to read and, if favourably impressed, would publish it in the *Annales de chimie et de physique* which he did in the following year.

Boussingault was never concerned about his own safety, for a journey through France at this time was fraught with danger. Travellers were mistrusted as he found when a party of dragoons intercepted him in the Savoie and wanted to remove him in custody to Chambéry under a law permitting the arrest of suspected persons, a direct result of the assassination of the duc de Berry in February of that year. However, he succeeded in convincing the dragoons that he was an innocent traveller and was allowed to continue his journey.[4]

He left Paris for Strasbourg, arriving late on December 18th, 1820 because a coach wheel had broken en route, near Chateau-Thierry. At Strasbourg he lodged with the manager of the Lobsann mine, M. Dournay, and was astonished to find that everybody spoke a language incomprehensible to him. The Dournays lived in Strasbourg during the cold Alsation winter and at this season Boussingault shared their Lobsann house with the mine accountant.[5] Unlike his letters from Saint-Etienne we now hear more about his *ménage.* His life at Lobsann was most pleasant, having servants and keeping a good table as the people of Alsace then had a better appreciation of food and good living than those of the Midi, as well as having a higher standard of literacy.[6] At least, these were his first impressions which he later modified, much to the delight of his sister Jeannette who believed that the French were superior to the Germans, and the Parisians the best of all peiple, the sentiment of one who appears never to have left Paris.[7] The Dournays' house was delightfully situated in the forest but a source of anxiety was the presence of a robber band composed of gypsies and army deserters from Baden on the opposite bank of the Rhine, and for this reason he kept two pistols handy at night.[8] He writes about his menus which included beef, mutton, German stew, veal and noodles, and his great delight, sauerkraut with pork; his table at Lobsann was superior to that at Saint-Etienne and cost less.[9] But he had servant problems, firstly because they spoke German, and he vainly attempted to entice his mother to Alsace to organise the household. A more serious matter was his dismissal of a Catholic servant because of her frequent attendance at Mass and Confession, and her replacement by a Protestant.[10] The indignant family in Paris protested that the servant had been dismissed because she was unsuitable and not because she was a Catholic.[11] Little is known of his work at Lobsann except that it was mainly underground and that three months were spent in tunnelling a gallery from opposite ends which met within a decimetre of error, a compliment to his surveying skill.[12]

Friendships with the Abbé Branthôme and the LeBels

Visits to Strasbourg led to friendship with the dean of the Faculty of Sciences at the University, the abbé Branthôme, who was also professor of chemistry. We have a clear conception of Branthôme's teaching because one of his students, Charles Schmidt, wrote up his lectures neatly in Indian ink and had them bound.[13] These notes form a signpost in the history of science because Branthôme, like most other chemists of his time, had rejected the phlogiston theory of combustion by which combustible materials were considered to contain a substance, phlogiston, which was released to the air on burning. In the simple case of a metal, it released phlogiston on burning leaving a calx – or what we would call an oxide. To reform the metal it was necessary to give the calx some phlogiston such as is contained in carbon or hydrogen – what was later regarded as the reduction of a metallic oxide. A number of chemists thought otherwise, especially Lavoisier who demonstrated the truth of the oxygen theory of combustion which could explain why metals gain in weight when burnt in air and which could desoribe the true nature of combustion and combustible bodies. But it could not account for the true nature of fire, heat and flame until, later, the vibratory nature of heat and light energy was understood. Lavoisier endeavoured to supply an answer by assumimg that all metals contain an invisible fluid, caloric, which was responsible for the temperature of a body, and if in excess it could explain the phenomenon of fire. In this manner the imaginary substance phlogiston became the equally imaginary material caloric. Thus, Branthôme could teach:

> If a hot body is surrounded by cold bodies it releases caloric to them and, because of that, it must make an effort to leave the hot body and enter the others in such a way as to produce an equilibrium between the hot body and those surrounding it. We call this effort Tension and is the cause of Temperature which we can define as the manifestation of the various energies of caloric.[14]

In the absence of the kinetic theory of matter, and the relationship between heat and energy, this was a most plausible explanation at the time. Branthôme's course was mainly on the metals and the non-metals, but it was completed by lessons on the plant world and included such topics as albumin, starch, gluten,

Fig. 3. Bechelbronn: the farm buildings. (Collection: M. Charles Boussingault.)

organic acids and fermentation. The syllabus bears some resemblance to that of Boussingault when, in 1845, he became professor of agriculture at the Conservatoire des Arts et Métiers in Paris.[15]

A more lasting friendship was with the family of Marie Joseph Achille Lebel of the farm Bechelbronn. It included his wife Marie, their fourteen year old son, Louis Frédéric Achille, then at boarding school at Strasbourg, and their six year old daughter Adèle, all living at the Château LeBel five kilometres distant from the mine at Lobsann. The LeBel family hailed from the Languedoc from where Antoine LeBel, the father of Marie Joseph, had migrated to exploit the oil-bearing sands of Pechelbronn. A concession of 10th November 1800 permitted the LeBel family to exploit an asphalt mine at nearby Lampertsloch, the official document being signed by Napoleon as First Consul, and the Minister of the Interior, Jean Chaptal, and which can be seen at the Musée de Pétrole at Merckwiller-Pechelbronn. Frédéric Achille, the son, became the father of Joseph Achille LeBel (1847–1930) the pioneer, with Jacobus van't Hoff of Holland, of the concept of stereoisomerism, which explains how two different forms of the same substance are due to different arrangements of atome in the molecule i.e., their different spatial arrangements.

Fig. 4. Pechelbronn anno 1760–1880. Until 1889 when liquid petroleum was first pumped to the surface, petroleum was distilled from bituminous sand in this building. (Collection: M. Charles Boussingault.)

The LeBel farm, Bechelbronn, was situated on the oilfield of Pechelbronn and erudite explanations have been offered to account for the spelling difference. A small river running with brown oil was known as the *Baechelbronner* or 'little brown stream' and gave its name to the farm which was known as *Baechelbronnen* in 1734 and *Bechelbronn* in 1821, undoubtedly the spelling adopted by the LeBel family who purchased the farm in 1768. *Pechelbronn* is a name associated with the brown pitch which is found locally.[16] Whatever the derivations it is convenient to assume that Bechelbronn refers to the farm and Pechelbronn relates to the oilfield. Today, a small portion of the original farm is known as *La ferme LeBel* on which a battery-hen industry is established and it also includes the Château LeBel, built in 1805 from the proceeds of the sale of oil to Napoleon's armies for the purpose of lighting, lubricating transport wagons and for treating wounds. The discovery of petroleum at Pechelbronn dates from the fifteenth century but it was not until 1863 that by drilling to fifty-seven metres, liquid petroleum was found and pumped to the surface in 1889. Prior to that, from 1740 onwards, bituminous sand had been excavated from hillside galleries and the petroleum extracted by distillation. Such enterprises were known as asphalt mines because of the bituminous material obtained from them.[17] Such was the background of Bechelbronn and the Château LeBel when Boussingault lived at Lobsann, but apart from the family who were most hospitable to him he made great use of the château library with its volumes of history and travel; his thoughts were again directed oversea.

His Meeting with Alexander von Humboldt and his Appointment to the School of Mines at Bogotá

His old urge to travel overseas was sorely tempted by a proposal from his former teacher and friend, Thibaud, now a mine manager at Grenoble, to join him in a most doubtful assignment in Egypt where two instructors were required for metal-casting and mineralogy. Thibaud appears to have had the impetuous character of a romantic and promptly accepted the posts for himself and Boussingault, yet nothing could be gleaned about conditions and prospects in Egypt, and perhaps for this reason and because of his pleasant and comfortable life in Alsace we hear nothing more of the project or of Thibaud.[18] Posts in Chile and Guatamals [19] were considered, which poses the question as to whether the desire to emigrate arose from the current economic instability of France, or because of his belief that geological study without overseas experience was both unsatisfactory and incomplete.

Fig. 5. Pechelbronn oilfield in the nineteen twenties. (Photograph presented to author by M. and Mme G. Kapp.)

About this time, in 1822, Simon Bolivar, the liberator of Venezuela and Columbia, decided to establish a training school for civil and military technologists at Bogotá, the capital of Columbia. Teaching staff were to be recruited by Bolivar's envoy in Europe, Francisco Zea, who was both politician and botanist. Earlier in his career he had been arrested by the Spanish authorities for sedition in Columbia and given a goal sentence in Madrid. He served his sentence by occupying the vacant chair of botany at the University of Madrid and the directorship of the Madrid botanic gardens! On returning to a liberated Columbia he became its vice-president in 1819. Zea had heard of Boussingault in a devious manner. Another Alsace friend of Boussingault was Phillipe Louis Voltz, a geologist and director of mines with whom he made frequent geological excursions. Voltz spoke highly of his friend to Pierre Berthier, director of the *Ecole des mines* in Paris, who, in turn, mentioned Boussingault to a Peruvian student of mining, Mariano Eduardo de Rivero. It appears that Rivero informed Zea, for Rivero had already received an appointment at the new institution in Bogotá, and in June 1822 Boussingault was appointed professor at the *Escuela nacional de mineres en Bogotá* at a salary of 7000 francs for a period of four years, a tremendous advance on his Lobsann salary of 1500–1800 francs. Until he left France in August 1822 he occupied a room in the rue Trainée near the Halles, a street that disappeared in 1845 when the rue Rambateau was constructed.[20]

Another intermediary was Alexander von Humboldt because of his previous experience gained during an exploratory journey through Venezuela and Columbia in 1804 in the company of the botanist Aimée Bonpland. Although a member of the Prussian court through family ties, Humboldt spent many years in Paris, living on the fourth floor of 26 quai de l'Ecole, now part of the quai du Louvre.[21] It is almost certain that he had replaced Zea as Bolivar's envoy because Zea fell ill and died in Bath in November 1822. Boussingault first met Humboldt when introduced to him by Rivero on the pont Neuf[22] and from that moment it was Humboldt who organised the expedition to South America for which he was not only eminently qualified but willing also as it was his greatest wish to return and establish a scientific centre, collecting meteorological and magnetic data on a large scale, and to train young scientists. Looking back over a period of sixty years Boussingault, in a vivid pen-picture, recalled Humboldt's assistance:

> Humboldt was actively interested in our expedition. We were not only travelling through regions which he had visited twenty years ago, but to live there and complete and extend many of the observations which he had made. The progress made in geology and geography since his memorable travels

> necessitated a close examination of the strata which he had too rapidly studied in passing, and of geographical positions which had not been measured with sufficient accuracy. . . . At first, Humboldt wanted to get to know me – to size me up. He talked a great deal and very well and I listened as a pupil to a teacher; he was pleased to observe that I had the 'great art of listening'. He showed much friendship towards me which he preserved until his death. He presented me with several instruments which he had used in America; a pocket sextant, an artificial horizon, a prismatic compass and a celestial planisphere of Flamsteed, valuable treasures of which I made the greatest use. . . . .
>
> Humboldt did even more. He insisted on teaching me how to use these instruments. . . . We measured the angle between the spire of the Invalides and the lightning conductor of Saint Sulpice church. We also found the sun's altitude. In my practical instruction he omitted nothing . . . and I was soon familiar with the use of the sextant and the artificial horizon. . . . [23]

In addition to this inestimable assistance Jean Fortin gave him two barometers and he received two fine chronometers from Abraham Breguet. The expedition had aroused the interest of Parisian scientific society and Boussingault met such outstanding scientists as Arago, Gay-Lussac for the second time, Laplace, Poisson and Biot, all interested in the subject of the barometric height at sea level at the equator. Previous readings, for example those of Humboldt (1799–1804) or those of Bouguer and Condamine (1755) when they were measuring a degree of the meridian in Peru, were suspect because their instruments had not been calibrated before leaving Europe. Boussingault's barometers and thermometers were calibrated at the *Observatoire Royale*, Paris, in the presence of Humboldt and Arago. Whilst preparing for the expedition Boussingault and Rivero spent several days in London at the Columbian Legation where they learnt that their boat was compelled to sail from Belgium or Holland owing to the hostility shown towards the rebel states of South America by the government of Louis XVIII of France.[24]

Of the party embarking for Venezuela, mining and geology were represented by Rivero and, of course, Boussingault; medicine and zoology came under the care of the physician-naturalist-artist François Roulin; Goudot was the botanist who later carried out researches on South American food plants. The mystery man was Bourdon, a former army physician, who was responsible for entomology. His record with the expedition was brief and unfavourable. Perhaps he was a friend of Audouin, a founder of French entomology, who attended the farewell dinner party at the chez Véry, in the arcades of the Palais Royal. Accompanying them to South America was the intrepid Mme Roulin and her four year old son Louis. The laboratory and library were dispatched to Cartagena in Columbia but Boussingault accompanied his instruments to Antwerp, where the travellers were to embark, as he and Rivero were to make magnetic and meteorological observations en route from La Guaira in Venezuela, their port of disembarkation, to Bogotá, whilst the others were proceeding by boat to Cartagena.[25]

In Antwerp not only did the sailing delay produce frayed tempers but the subsistance allowance became exhausted. Humboldt in Paris was distressed to hear of these disagreements and promptly sent Boussingault a thousand francs with a typical Humboldtian gesture – "Don't hurry to repay me. You shall give it to me one day in Mexico in platinum and palladium". But he gently reproved Boussingault for failing to mention his fellow travellers in his letters, not even to inform him that Roulin was suffering from a fever, for the introvert was once again asserting itself.[26]

That Boussingault could be sent as a scientist to South America without specific scientific training was a tribute to him but also to the flexibility of French scientific and technical education and to the openness of academic recruiting. Generally, candidates were not necessarily selected from those with degrees and diplomas, or came from certain prestigious institutions.[27] How else can it be explained that of a scientific expedition the personnel were largely, if not wholly, amateurs. Public lectures, an elementary mining school as Saint-Etienne purported to be, and professional friendships with people such as Thibaud, Beaunier, Voltz, Arago and Humboldt had produced a young man of twenty years who was prepared and eager to solve the scientific problems of a relatively unknown sub-continent. Perhaps we must admit that he was over-eager considering the flimsy nature of his scientific background. His motives were clear. Generally, he believed that for a profound understanding of geology some overseas experience was necessary: specifically he was attracted to South America for there he could exchange the dead volcanoes of his native Auvergne for the active giants of the Andes.[28] These were the influences and the incentives which explained his presence on the deck of the 'New York' about to sail on September 22nd from Antwerp harbour for a new world and a new life.

CHAPTER 3

SOUTH AMERICA

From Antwerp to Dover the 'New York' flew the flag of the United States but henceforth it was replaced by the Colombian national flag and the vessel became the 'Patriot'. Such a routine change might have passed with little comment but for the transference of arms and ammunition from another vessel – the 'New York', alias 'Patriot', was truly a Colombian warship with all the risks and dangers attached to a rebel man-of-war. The possible consequences of such action were clear to the crew members who deserted at Portsmouth. With a new crew, the route followed the islands of Madeira, Barbados and Tobago, but the voyage was far from peaceful. Early in the voyage the crew of a Spanish ship appeared ready to board the 'Patriot'. The alarm was accompanied by the distribution of cutlasses to all including Boussingault who afterwards claimed that this diversion cured him of an attack of sea-sickness! More exciting was an exchange of gunfire between the 'Patriot' and a Spanish frigate, the 'Maria Francesca' which was captured and taken as a prize to the Venezuelan port of La Guaira.[1]

Fearing an attack by a Spanish warship, the Roulins, Bourdon and Goudot decided to leave the 'Patriot' and proceed in a lugger with square sails which had been purchased by Roulin, sailing as far as Barranquilla at the mouth of the rio Magdalena, the gateway to Columbia. Up to Magdalena they were punted in a sampan-like craft and disembarked at Nare from whence they travelled by mule to Bogotá, arriving in March 1823,[2] two months ahead of Boussingault and Rivero who had disembarked at La Guaira for an exploratory journey overland on horseback to Bogotá. The route taken by Roulin's party was by this time well-known as British government and trade representatives, anxious to establish themselves in Colombia, were using it as it was the only possible route to Bogotá.[3]

Boussingault and Rivero Leave for Bogotá

Boussingault arrived in Venezuela at a time of political and economic turmoil throughout the sub-continent. The colonies of South America had revolted against Spanish rule in 1810. Starting as an opposition to the French control

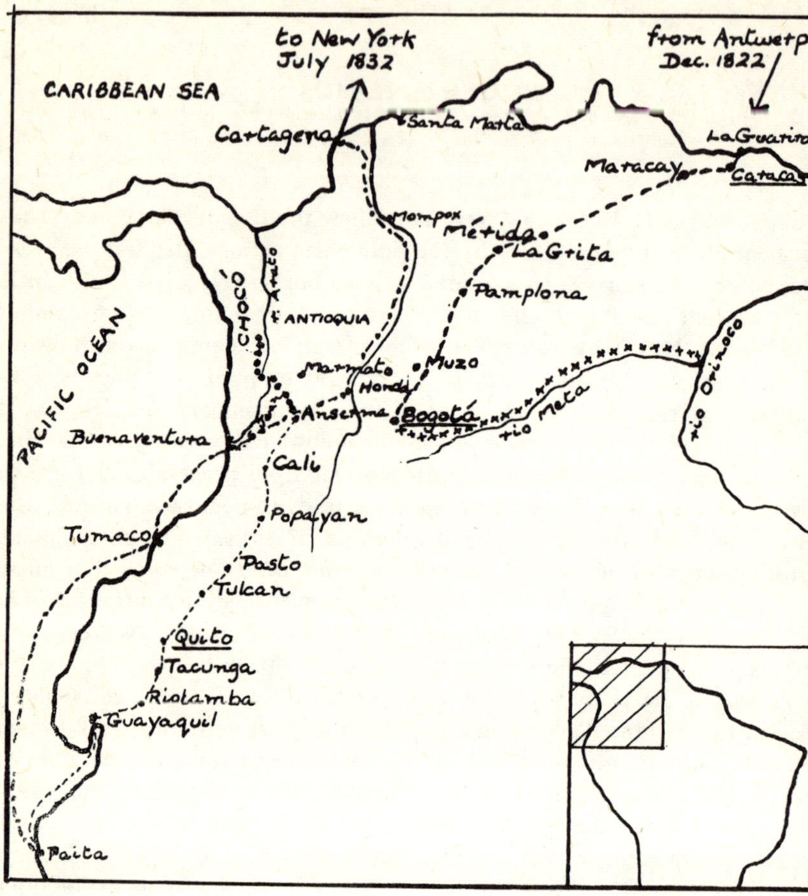

Caracas to Bogotá 1823 – – – –
rio Meta 1924 x x x x

Choco 1829 • • • •
Andes 1830–31 ....
To Cartagena 1832 –·–·–

Fig. 6. Boussingault's journeys in South America (1822–1832).

of Spain when Napoleon's brother, Joseph Bonaparte, supplanted Ferdinand VII in 1808, it became a struggle for independence from Spain. However, revolutions in Venezuela in 1814 and in Nueva Granada, later Colombia, in 1816 were failures due to personal separatism which hampered the liberation movement. In Venezuela, creoles born in South America, were against the

Fig. 7. Boussingault in 1822, drawn by Dr F. Roulin *en route* to South America. (Collection: Mme F. Boussingault.)

peninsulares who were born in Spain; in Nueva Granada the creoles opposed the fast riding llaneros of the plains who were of mixed Indian and Spanish stock. Simon Bolivar, who had played a prominent rôle in the abortive revolution in Venezuela, succeeded in August 1819 in defeating the Spanish army at Boyacá in Nueva Granada and proclaimed a union of Venezuela and Nueva Granada in the Republic of Colombia. The new republic received its constitution in 1821 with Bolivar as its first president. The presidency of Quito, later renamed Ecuador, became part of Columbia in 1822, and Guayaquil with its port on the Pacific Ocean was annexed in the same year. Venezuela was now free from Spanish control except for the port of Puerto Cabello on the Caribbean Sea. Bolivar's victory at Ayacucho freed Peru, followed by Upper Peru which was renamed Bolivia in 1825 as a tribute to its liberator. Bolivar had now reached the height of his power and influence and proposed an Andean Confederation at the Congress of Panama in 1826. The Confederation failed because nationalism overcame internationalism in spite of Bolivar's plea for unity. Weak administration and the dictatorial character of Bolivar were also contributing factors. Even his Colombian Union broke up into the three independent states of Venezuela, Nueva Granada and Ecuador under rival generals. Exhausted and ill, he was the object of an assassination attempt and so he decided to recuperate abroad, but at the port of Santa Marta he died from pulmonary tuberculosis in December 1830 at the age of forty-seven. His dream of a Confederation was in ruins but today it is recognised that he was mainly responsible for the liberation from Spanish rule of much of South America.[4]

Threatened by the prospect of capture by General Morillo's Spanish troops Boussingault and Rivero postponed their departure for the interior and started a series of barometric observations which confirmed that the barometric height at sea level in the tropics was the same as in temperate regions.[5] After which they travelled on the road to Bogotá with a detachment of Bolivar's army which was a wise precaution because the remaining Spanich armies had the unpleasant custom of cutting the throats of their captives, especially those who were in the employ of the rebel Colombian government.[6]

The Cause and Cure of Goitre

Halting at La Grita in Venezuela the officers of the accompanying troops organised a ball to which the young ladies of the town were invited. In his *Mémoires* Boussingault could not conceal his stupendous amazement on beholding the large goitres characteristic, without exception, of all the ladies present. He gave dimensions of the largest goitres which we trust were estimates only, and reported having seen a goitrous dog in the town.[7]

Boussingault's hypotheses on the cause of goitre could only be tentative and pragmatic but he appears to have made a distinction between cause and treatment. His own observations compelled him to dismiss various theories of its origin which attributed the disease to excessive consumption of alcohol, lack of personal hygiene, the nature of the surrounding rocks, coarse food, drinking water obtained from melting snow – because of its low temperature or because of dissolved salts – and a humid climate. This last theory was current in France in the Vosges and the Pyrenees, areas of heavy rainfall.

As is well known, goitre manifests itself by a swelling of the thyroid gland situated in the neck. There is a long history of observations on the thyroid gland. The Greek physician Galen, living in the second century A.D., believed that it was the source of fluid for the lubrication of the larynx. Thomas Wharton (1614–1673) a London physician, gave the gland its name from the Greek *thyreos*, a shield, for such is its shape, and thought that it was designed by nature to give, especially in females, a rotundity and beauty to the neck. The rich supply of blood vessels in the gland gave rise to the early nineteenth-century hypothesis that its function was to protect the brain from the pressure of the circulation. It was not until 1896 that the presence of an iodine compound, thyroxine, was recognised, then isolated in 1915 and synthesised in 1927. Thyroxine helps to regulate various aspects of body metabolism, e.g. pulse rate, strength of heart beat, nerve response to stimuli, and rate of body growth, so that absence or insufficiency of thyroxine causes an enlargement of the gland in an effort to produce more of this amino-acid. As far back as the twelfth century the ash of seaweed and sponge – both sources of iodine – was recommended as a treatment.[8]

Although there was a time when he suggested that dissolved salts in the drinking water was a possible cause, Boussingault appears not to have read the work on the subject of Francisco de Caldas, an outstanding Colombian scientist who was executed by the Spanish troops for his part in the anti-colonial struggle. Caldas is claimed to be the first, in 1808, to attribute the disease to excessive 'lime' in drinking water. Also in Colombia Don Joaquim Camacho in 1810 found that the disease was rife in the convents of Bogotá where the wells supplied hard water,[9] a reminder of the presence of 'Derbyshire neck' in England. Another century was to pass before the discovery of goitrogens, substances that inhibit the synthesis of thyroxine and therefore its subsequent passage to the body. A significant observation of Boussingault was that some Colombian families avoided goitre by obtaining their water supply from a distant source and thus suggested that certain water sources might contain a harmful substance which encouraged a goitrous condition which led to him advocating the use of rain water.[10] Only recently has it been realised that this was an extremely important observation as endemic goitre is now believed to be due not only to nutritional iodine deficiency but to the presence of organic compounds containing sulphur, found in the water supply. These organic compounds are found in sedimentary rocks and their presence restricts the production of thyroxine. In Boussingault's day calcium compounds were suspect and as these account for much of the hardness of water it is apparent that Caldas, Boussingault and Camacho were pioneers on a trail which could not then be clearly defined until the structure and the function of the thyroid gland had been determined. The concept of a goitrogen was due to McCarrison whilst in India in 1906, when a situation was presented to him which resembled that of Boussingault when noting that goitre-free families obtained their water from a far source.[11] McCarrison included himself in a test team which developed progressively increasing goitres. Suspected in 1906, the presence of goitrogens was confirmed in rabbits in 1928.

As were many others, Boussingault was impressed by the prevalence of the disease in communities living above an altitude of 2000 metres and assumed that the 'thinner' air at that height provided insufficient dissolved air in the water supply. He cited the case of a family which avoided goitre by leaving their supply in a tank for 30 to 40 hours in order to aerate it, as he thought, but perhaps there was in that time a sedimentation of goitrogenous substances.[12]

If Boussingault was uncertain of the cause of goitre he was on surer ground when discussing the cure. Already mentioned was the ash of seaweed and sponge which the Swiss chemist Coindet of Geneva found, in 1820, to be due to the presence of iodine, an element which had only been discovered nine years previously. In the same year Coindet successfully administered tincture of iodine to goitrous patients but the method was later abandoned because of its toxic effects. An improved treatment had already appeared in 1816 when William Prout, outstanding physician and nutritionist, gave his patients the salt potassium iodide, after testing it on himself to verify that small doses were not toxic. Prout mentioned that this remedy was in use in St. Thomas' Hospital, London, in 1819. Therefore, Boussingault was probably aware of the prophylactic effect of iodine on goitrous conditions and noted that in goitre-free communities the streams contained iodine in the form of iodides,[13] yet it did not occur to him that goitre could be due to the absence of iodine in water used for culinary purposes. However, he strongly advocated the use of iodine in diets of goitrous populations and that iodised salt should be generally available, yet his recommendation of 1831 was not implemented by the Colombian government until 1947 as a result of the disclosure that 10 per cent of prospective army recruits in 1942 were rejected because of simple goitre.[14]

During Boussingault's sojourn in South America the iodised salt which was then used was a by-product of silver mining. Common salt, required for the extraction of silver from its ores, was obtained by the evaporation of water from salt streams or marshes, carried out in copper boilers. The saturated salt solution thus obtained was filtered through inverted earthenware cones in which the common salt crystallised, whilst an oily filtrate, containing among other salts various iodides, passed through the earthenware filter. This oily liquid known as *aceyte de sal* had a ready sale in the goitrous regions of Colombia.[15]

As would be expected, there were cretins, those suffering from a form of goitre associated with symptoms arising in early infancy. Boussingault narrowly escaped assassination in a market place by a cretin who was just prevented from stabbing him in the back. In spite of this unnerving incident, or perhaps because of it, he expressed great sympathy for these unfortunates who were not only venerated but in the eyes of their community could do no wrong. Some attained high office for Boussingault wrote of meeting them in parliament, in professional societies, at court and in the civil service.[16]

Some Interesting Minerals

The salt-water lake of San Juan was on their route, near Mérida in Venezuela, and this provided a local industry [17] as from its bed was obtained a carbonate of sodium, 'urao', used by the Indians when chewing tobacco in order to increase the flow of saliva. The bed of the lake consisted of a top layer of black mud under which was a yellow clay. The underwater 'mine' was pegged by five metre long posts driven into the bed. Three men at a time worked the mine, one in a boat who held a pole slantwards towards the mine whilst the other two jumped into the lake and made their submerged way guided by the pole. After a minute they would reappear with handfuls of 'clavos' and 'urao'. The sharp, needle-shaped crystals were 'clavos' or 'nails' and were found in the yellow clay. Upon analysis Boussingault found them to consist of sodium calcium carbonate,[18] a hitherto unknown mineral to which he and Rivero gave the name 'Gay-Lussite' in honour of the famous Fernch chemist. Beneath the yellow clay was a layer of *urao* which is a sesquicarbonate of sodium. In his *Mémoires*, looking back over a period of sixty years, Boussingault regarded the 20th April, 1823 as a particularly good day as he had discovered two new minerals, but he seemed to have forgotten that his collaborator in these discoveries was Rivero.[19]

Just four days to go before reaching Bogotá they noticed at Santa Rosa a blacksmith's anvil of meteroritic iron which had been found by a small girl on Holy Saturday, the Saturday before Easter, in 1810. As it was found on a well frequented path there was no doubt about when the meteorite fell, and its impression on a superstitious people who generally regard a meteorite as a piece of the solid vault of sky, or associated with thunder, lightning and the falling of stars, or more generally, a symbol of inexplicable and stupendous preternatural power. The mayor of Santa Rosa kept it for ten years and then the local blacksmith took possession but it was always understood that the anvil was the property of the girl, who, now in her early twenties, was induced to part with it to Boussingault for a satisfactory sum. He intended it for the museum at Bogotá but the news of the sale brought forth several other pieces from which he forged a sword, and with youthful enthusiasm he had it engraved, 'Iron sent by God for the defence of liberty'.[20] Had he read that the Bedouin of Sinai believed that a man who succeeded in forging a sword of meteoric iron was invulnerable in battle?

The 24th of May, 1823 saw them at Bogotá at last, having covered 1525 kilometres on horseback. Boussingault was immensely impressed by the country around Bogotá which, throughout the year, enjoyed the climate of a European Spring. The indigenous population, Muyscas, were agriculturists but kept no cattle, a testimonial to the available nutriment in the soil. He observed that they used a wooden plough, grew maize, potatoes, arracacha – a parsnip-like vegetable – and quinoia, the seeds of which were ground into a flour and made into cakes, or boiled like rice.[21] It appears that the opening of Bolivar's school of mines was delayed for hardly had he arrived in Bogotá when he was asked to inspect gold reefs in the Magdalena valley between Ibagué and Honda, an assignment in a hot district which occupied two months.[22] Whilst in the Magdalena valley he attempted an investigation of a topic which appears to have received little attention, that is the variation in the intensity of sounds between day and night. He was camped near the Guali river, west of Honda, when it was a raging torrent, and at various times during the day and night he made accurate readings of temperature and humidity to discover whether there was a correlation between these factors and the intensity of sound from the river. Perhaps he had yet to learn that if his readings of temperature and humidity were precise then it was useless to adopt such subjective values of sound intensity as weak, loud, very loud and uncomfortably loud. No general conclusions could be drawn from these readings as there was no correlation between the intensity of sound and the temperature or the humidity. He attributed the difference between day and night sounds to the upward currents of air occasioned by the solar heating of the earth.[23]

Malaria and the Meta Expedition

Next came a highly dangerous assignment, an expedition along the banks of the rio Meta up to its confluence with the rio Orinoco, through malarial country.[24] The symptoms of malaria were described either by Hippocrates or one of his students in the fourth or fifth century B. C. It was long known as the ague in Britain and the fièvre in France but the term 'malaria' did not appear until the middle of the eighteenth century and then did not specifically refer to the disease until the latter half of the nineteenth century. The discovery of the parasitic protozoon responsible for malaria was due to a military surgeon, Charles Lavaran, in 1880 whilst with the French army in Algeria. As a result of his discovery in 1894 that elephantiasis in man is conveyed by a Culicine mosquito, Patrick Manson advanced a theory that a mosquito was the second host of the malarial parasite and this was confirmed by Ronald Ross of the Indian Medical Service in 1897 who saw stages of the protozoon life-cycle in the stomach wall and in the salivary glands of Anopheline mosquitoes. The general life-cycle of the protozoon, Plasmodium, was elucidated in 1900 by three Italian scientists, Bignami, Bastianelli and Grassi. This brief inclusion of facts well known in the history of medicine shows that the origin of malaria was unknown in the lifetime of Boussingault and therefore the suggested remedies had little relation to the cause of the disease. Before departing for South America he had observed the effects of malaria in the Ain département of eastern France where, outside Bourg-en-Bresse, locally grown flax was steeped in the pools of the neightbouring marshes. Boussingault noted the unhealthy appearance of the peasants suffering from malaria.[25] He was, himself, the victim of a severe attack when on the rio Meta expedition, early in 1824 in company with Rivero and Roulin and accompanied by a picket of soldiers for protection against the Indians. The object of the expedition was to survey the course of the Meta and to make observations of latitude and longitude of various places including the confluence with the Orinoco.[26] Humboldt was anxious to know of this last reading because at the beginning of the century he had found the longitude by chronometer observations but failed to find the latitude owing to a perpetually cloudy sky, especially at night when it was essential to see the stars. The Meta flows through vast tropical plains populated mainly by Indians and the half-caste creole-Indian horsemen, the llaneros, who live by ranching. It is known that the party reached their objective because in a letter referring to the position of the confluence Humboldt was gently oritical of Boussingault's methods of surveying, his manner of fixing the stars and the absence of data concerning his chronometer.[27]

During the expedition the party was pestered by mosquitoes, a cloud of them descending on a ham which was part of a riverside lunch. Boussingault held the ham above his head and rode the horse for nearly two kilometres but failed to elude the mosquitoes. He was advised to wear a sheepskin coat, the odour no doubt discouraging the insects, but he found the coat too hot. Although cures had been effected with cinchona bark since the middle of the seventeenth century he had little faith in it for a permanent cure. When desperately ill he was treated at a mission, Giromira, by the church sexton who gave him a medallion with the effigy of Our Lady stamped on it. These medallions were prepared at a Capuchin mission on the Orinoco by mixing powdered cinchona bark with clay and moulding them. For the treatment the medallion was powdered and boiling water added; when sufficiently cool the mixture was imbibed by thc patient. Boussingault was also subjected to a starvation diet following the then popular precepts of François Broussais (1772–1838), who taught that most illnesses were the result of tissue irritability caused by over-stimulation through a generous diet. A corporal from the picket was ordered to accompany him back to Bogotá where he had a relapse but recovered with the help of the corporal who, scorning the Broussais regimen, fed him with meat which he had pre-masticated during the night. But before his recovery, whilst in a malarial coma, he noticed Bourdon, the physician-entomologist of the party, in the act of removing some of his silver cutlery when he was interrupted with, "Bourdon, I am not yet dead." Boussingault excused him, as being a former army doctor he had probably formed the habit of robbing the dead![28]

Depending on the situation and the time, Boussingault favoured various theories of the cause of malaria. They included excessive humidity, large temperature differences between night and day, and excessive temperature differences between the body extremities as in the case of Cornish miners who were recruited for work in Colombia. Their heads were exposed to the tropical sun whilst their feet were in the icy streams from the Andes, and eventually they were replaced by the local Indians. Finally, he was convinced that there was a harmful constituent of the air which caused fevers, a miasma which he sought to identify.

It was Giovanni Lancisi, physician to the Pope, who endeavoured to explain in 1717 that outbreaks of disease were due to miasmas of the atmosphere. His miasmatic theory was published in a work which can be translated as *On the Noxious Effluvia of Marshes*. Miasmas were believed to arise from putrid plants and animal remains, from drains and cesspools, and were the cause of sickness. The miasmatists were opposed by the contagionists who believed that disease was contracted by contact with another person who was suffering from that disease, and with the rise of the science of bacteriology in the nineteenth century the contagionists were able to refute the claims of the miasmatists. Boussingault believed that miasmas arose from the effect of heat and humidity on decomposing plant and animal material, but suggested, in addition, that they could arise from felled trees as he found that afforested regions were unusually malarial. He envisaged a war between the forces of putrefaction and man, which in the tropics was a continuous struggle but that in the temperate regions a truce was declared in winter. He made certain *a priori* assumptions about miasmas and attempted to show that 'bad air' was due to the presence of a flocculent hydrogen compound against which the inhabitants took precautions by covering the mouth whilst crossing marshy places. At Cartago, on the rio Cauca, he conducted experiments to verify his assumptions when the south wind blew in from the marshes and appeared to cause illness in the town. 'Unhealthy' air was dried by its passage through the usual drying agents, calcium chloride and sulphuric acid, and then passed through a red-hot glass tube where any hydrogen compound would be exidised to water and collected in a weighed calcium chloride tube. The results of the experiments were not at all convincing, for although some water was formed, indicating the presence of a hydrogen compound, there was a suspicious fluctuation in the results because less water was collected when the humidity of the air fell, suggesting that the air had not been completely dried. On his return to Paris this experiment was repeated at his brother-in-law Vaudet's home in the Marais in April and May 1834, and again in Lyon in 1835 where he was dean of the Faculty of Sciences at the Académie de Lyon, but it appeared that a hydrogen compound in minute quantities was a characteristic of city air.[29] However, he was in good company because Justus von Liebig (1803–1873) also believed in miasmas as the cause of disease but in addition that they communicated their state of decomposition to the blood. Unhealthy blood, according to Liebig. emitted gaseous contagious matter as a miasma capable of generating itself again in other blood. Substituting 'bacteria' for 'miasma' we see that Liebig was close to the modern theory of contagious disease.[30]

Curare

During the Meta expedition, the travellers saw Indians killing birds with arrows tipped with curare and propelled from a blow-pipe 1½ metres long, the bird usually succumbing in five minutes. On one such occasion, when a bird had been killed in this manner, Boussingault was the centre of an extraordinary incident which he described as follows:

Then the Indian, who was looking at me whilst holding his blow-pipe vertically, gave me to understand by a haughty gesture that I should pick up the bird. For a moment I admired his pose and the beauty of his figure, then, considering that he lacked respect I gave him a resounding smack on his buttocks. It is too easy to spank the trouserless. With a meaningful gesture, not as well executed as his, I called upon him to pick up the bird which he did, walking slowly with grace and dignity.[31]

One wonders whether Boussingault would have 'lost face' had he picked up the bird himself?

The use of curare was first described in 1516 by an Italian, Ascoria Sforza,[32] who published the letters of his New World friend, Peter Martyr d'Anghera. Humboldt wrote it as *uiraery* which is said to resemble the Indian word and is probably correct as he had a keen ear for languages; was not his brother Wilhelm a renowned philologist? Unlike our travellers in the Meta valley, Humboldt and Bonpland saw the preparation of curare by the evaporation of the sap of various plants, and although they discovered the Indian names they did not give the Latin equivalents.[33] Boussingault mentioned that the epileptic General Paez, who became president of Venezuela after the break-up of Bolivar's Colombian Union, took curare orally when attacks were imminent, and in such large doses that would have proved fatal had they been taken intravenously.[34] The interest aroused in Roulin and Boussingault over the subject of curare led them to make a chemical examination of the substance. They disproved the current belief that it was a form of strychnine which had been recently discovered in nux vomica, the seed of an East Indian tree. They concluded that it was a vegetable alkali but, like other workers on the subject, they were frustrated by the extreme variability of crude curares when endeavouring to isolate the active principle, because on exposure to air it is hygroscopic and becomes a yellow or brown syrupy liquid.[35] The investigation of curare was related to the discovery of two different species of sting-ray by Roulin, for the Indians of the Meta used the sting, coated with curare, as a tip to their arrows. Until they were harpooned in the rio Meta the sting-ray was considered to be a sea fish.[36]

Boussingault's Varied Posts

Like many posts in the early colonial service of most imperial powers, that of Boussingault was clearly defined on appointment but in practice was not so clearly demarcated. He was appointed as a professor of a new school in Bogotá founded by Bolivar for the purpose of training technologists for civilian and military needs.[37] Whether the school achieved its ambitious aims is doubtful and we know practically nothing of his programme there except that he taught chemistry but found it difficult because he was obliged to teach in Spanish.[38] He was probably teaching in 1823 and he was certainly teaching in 1824 but thereafter there is no mention of his appointment at the school. In practice, Boussingault's post became a combination of teaching, exploration, surveying, prospecting, and the inspection and management of mines which involved an extensive programme of assaying, using methods which he modified when new problems arose in the analysis of ores. The occasional pacification of protesting or striking miners undoubtedly called for a high degree of diplomacy from this young man.[39] The contract with the Colombian government was for four years, 1823–26, after which he expected to return to France, but in 1827 he accepted employment with the Colombian Mining Company[40] registered in London, although still retained by the government to reconcile its interests with those of the company, most probably a form of public relations. The Colombian Mining Company was intent on discovering and exploiting the gold mines formerly worked by the Incas which had provoked the imperialist aims of Spain in the sixteenth century. Throughout the ten years in South America he held an army commission and left, in 1832, with the rank of lieutenant colonel although he was never occupied with military duties.[41]

The first mention of a mine inspection other than a gold mine is a visit to the rock-salt mines of Zipaquira, north of Bogotá where a nationalised industry had caused a village to grow to a community of five thousand, all engaged in the salt trade. Frauds and embezzlements had been committed and there were some 'black market' dealings by which the government had incurred losses. Because a previous military detachment had been disarmed by the angry populace and a soldier killed, Boussingault was preceded by a company of artillery! The mission was successful because not only did he stay in Zipaquira but completed a study of the rock-salt reefs in the neighbourhood.[42] The emerald mines of Muzo, also north of Bogotá, were visited about the same time.[43] Emeralds were found in rock debris, the result of disintegration of schist by the action of water, smaller emeralds being found in the fields and even in the gizzards of fowls. The emerald mines, as well as a number of farms, were owned by Pepé Paris, a man of great energy and initiative. There was a time when Boussingault thought he would marry his daughter, Manuelita. Apart from his marriage in 1835 there is no metion of friendships with women other than with Manuelita which was to have culminated with an engagement party. Buoyed with hope and excitement he gradually realised that no such announcement was to be made; it was the end of the party and the end of his romance. Manuelita had no intention of spending the rest of her life in Europe.[44]

Gold Mining

Boussingault's period of employment with the Colombian government was one of variety, not only in the districts visited but in the minerals examined. As a contrast his new employers, the Colombian Mining Company, were limited in their prospecting by concessions in the districts of Antioquia and Vega de Supia, and to the extraction of gold with its accompanying silver and platinum ores. Gold extraction followed the methods of the Incas and he noted the various stages which differed slightly according to whether the ore was mined from a gallery in the side of a hill or whether it was obtained from alluvium washed down the rivers, especially after heavy rain. In the former type of mining, the galleries in the hillside opened on a stream in which the ore was washed. The ore was dug not with picks as in Europe but with the more effective *vara*, an iron bar pointed at one end and with a wedge at the other. The ore was carried to the stream in leather bags and then washed. Women then pulverised the clean ore with a pestle and it was then transferred to a *batea*, a conical cedar dish, so that the heavy gold particles were found in the centre of the dish. Another *batea* washing, or panning, was necessary as the gold from the first washing contained a black sand consisting of ferrous titanate. The other method was practised on the flat llanos where the muddy river water, especially after storms, was directed into trenches and the debris

Fig. 8. *En route* for the gold mines of la Vega de Supia: Boussingault is seated on the right. (Drawing by Dr F. Roulin from *Roulin et ses amis 1796–1874* by Marguerite Combes (éditions Victor Attinger, 1928).)

removed by the speed of the current. The black sand at the bottom of the trench was submitted to *batea* washing by negresses.[45] In either method the grains of gold were melted into ingots.[46] Boussingault was highly critical of the gold extraction methods because of the wastage of ore in the trenches and streams which resulted in much gold being lost in the river. It was said that nobody asked for charity – they found it in the river.[47]

The Platiniferous Region of the Chocó

In a letter to his mother in 1823 Boussingault expressed the hope that he would be sent to the platiniferous region of the Chocó which is the north-western portion of Colombia with a seaboard on the Pacific Ocean.[48] The presence of platinum in this region was first reported by two Spanish naval officers, Jorge Juan and Antonio de Ulloa who travelled to South America in

1736, although rumours of a mysterious metal, hard and infusible, were known in the sixteenth century and it was said to be found in Mexican gold mines. Juan and Ulloa accompanied the French scientists Bouger and Condamine who were sent to Peru to measure a degree of the meridian. South America was then closed to foreign visitors by the Spanish government and the presence of the two Spaniards was to demonstrate that the king of Spain approved of the expedition. Juan and Ulloa provided the first description in Europe of alluvial gold-mining in the Chocó, and, indeed, this was the first literature on platinum to find its way to Europe. They described the difficulties of separating gold from the adhering platinum for which there was little use except as small shot, or for small bags which when filled were used as clock weights, or even for making garden paths. Otherwise it was thrown away and mines whose ores contained excessive amounts of platinum were often compelled to close down.[49]

The platiniferous area of the Chocó is situated in the river basins of the rio Atrato and the rio San Juan, the Atrato flowing north to the Caribbean Sea and the San Juan flowing south to the Pacific Ocean. Their sources were so close that Boussingault mentions the work of a priest at the gold-refining town of Novita who, in 1788, encouraged Indians to dig a canal connecting the two rivers and thus providing a link for small craft between the Pacific and Atlantic Oceans.[50] The Chocó at this time was a malarial area of almost continuous rainfall, plagued by mosquitoes and known for its impenetrable forests where communication was only possible by river. The alluvial deposits, owned by Spaniards, were worked by negro slaves whose rations consisted of salted meat, dried bananas and maize biscuits brought from the fertile Cauca valley on the backs of porters, for attempts at crop cultivation and ranching had failed.

Boussingault's Journey in the Chocó

Boussingault left Anserma, in the department of Caldas, in February 1829 accompanied by the botanist Goudot and several bearers. The Spaniards appear to have maintained some degree of social life for at the first town, Tamana, they attended a ball at which an uninvited guest was a jaguar. To Novita they travelled down the rio Tamana in hollowed-out tree trunks and from there they journeyed to Tado, the centre of the platiniferous region. There, the priest informed the incredulous Boussingault that platinum could be found locally, mixed with an insignificant amount of gold. He was taken to the priest's garden where the negress was washing the garden soil in a *batea*, a wooden dish used for panning, during which the heavier gold and platinum particles sank to the bottom of the *batea*. The negress was obtaining platinum with only a few grains of gold, for the garden had been a washing place where platinum had been removed from what was then the more precious gold.[51] Further from the rivers platinum and gold-bearing ores were crushed and washed with water, after which the powdered ore underwent sedimentation in trenches which effected a rough separation of the precious metals, the final separation being accomplished in a *batea*. Deplatinisation of the gold was then carried out at Novita by use of mercury which forms an amalgam with gold but not with platinum, and finally the deplatinised gold was sent to the mint at Bogotá.[52]

The origin of the alluvial platinum of the Chocó had been a subject of speculation since 1785 when it was observed that whereas only fine particles of platinum were to be found in the plains, larger particles were found nearer the mountains. With a hint from Humboldt, who knew the country as a result of his explorations of 1799–1804, Boussingault discovered the source of the alluvial platinum in a reef which he studied at Santa Rosa in the Cauca valley of Antioquia in 1825 where, in the Cordilleras, he found grains of gold and platinum in the rock syenite.[53]

The Bolivar Statue: A Lesson in Diplomacy

The discovery of the platinum reef coincided with the zenith of Bolivar's career in 1825 and this may have been the reason for the Colombian Congress voting for an equestrian statue of Bolivar in platinum; but let Boussingault tell the story:

I was at Bogotá busy with the map of Colombia when Congress decided to erect in the Plaza Mayor of the capital an equestrian statue of General Bolivar in platinum as a lasting tribute to the man to whom they owed their liberty. Several days afterwards I received from the Ministers of War and Finance an official letter appointing me to superintend the casting and the erection of an equestrian statue in platinum of the Liberator. The letter was received through the official channels, i.e., through Colonel Lanz. I had to reply to the Minister of Finance. I replied politely that I could not undertake the work which had been entrusted to me; that the necessary quantity of platinum was so great (I stated the weight) that all the mines in Colombia could not produce it in the course of a century; concluding where I should have started, by saying that since platinum was infusible by the usual technical processes it would not be possible to cast a statue in this metal.

Lanz said to me that all this was correct but in view of my position my letter lacked common sense because it proved that Congress and the Ministers were ignorant which was inexcusable even if I was right. Write this, he added, I shall dictate the reply which you should make.

I thanked the Minister for wishing to entrust me with so important a work, adding that I would spare no effort to assure its success. Before signing I again told Colonel Lanz that success was impossible since the fusion of platinum was impracticable. It doesn't matter, he replied, you have promised to make every effort. Besides, you know that you will never have enough metal. It will be forgotten and you have offended nobody. It happened as Lanz had said; the Minister was delighted and thanked me for the enthusiasm which I had shown, then it was forgotten. Altogether I received two kilograms of platinum which was used to construct several pieces of apparatus in the laboratory.[54]

If these few pages appear to do scant justice to eight or nine years of scientific labour it is because his curiosity ranged over a wide field. He investigated the obscure chemistry of the Patio process for extracting silver from its ores, and the amalgamation process for separating silver from the gold and silver alloy produced at the mines. The body paints of the Indians in the Meta valley were of particular analytical interest but their significance was unknown to him; their symbolic use was never studied until a century later when A. R. Radcliffe-Brown, an anthropologist, sought their meaning among the peoples of the Andaman islands. Again, we should not dismiss the extensive list of mineral analyses, conducted mainly at Marmato, or his analyses of thermal springs emanating from volcanoes, and his observations of earthquakes. Yet authorities in France and Britain have criticised the paucity of geographical information accumulated during his sojourn in South America, forgetting that, unlike Humboldt, he was not a free agent or a full-time explorer but an employee of either the Colombian government or the Colombian Mining Company.[55] Many of the observations made during these years and much of the research undertaken were to sustain his literary output for the rest of his life. Even at the age of eighty-one he produced memoirs on such diverse subjects as the cultivation of cocoa and the bronze tools used by the Aztec miners of Peru.

But France was now calling; his parents had not seen him for nearly ten years and during that time he had been absent from the centres of research, and above all from Paris. Yet the volcanoes of the Andes remained to be explored.

CHAPTER 4

THE GREAT VOLCANOES OF THE ANDES

Boussingault was in the gold-mining district of la Vega de Supia when he heard of Bolivar's death in December 1830. Disillusioned by the chaotic political and economic state of the country following the disintegration of the Colombian Union, he had already decided to return to France in spite of Bolivar's offer of the directorship of the Bogotá school of mines. Boussingault attributed the failure of the Colombian Union to the personal defects of Bolivar and to the weaknesses inherent in the population. He admired Bolivar as an honest and persevering man of simple tastes and sober habits, although he admitted his fondness for women. He was quickly enraged but calmed down just as rapidly. He described Bolivar as one who had not the strength of character to accept his temporarily elevated position of Liberator and he believed that Bolivar recognised this weakness since he avoided diplomatic contacts for fear that his public image would be shattered by the reality. Boussingault regarded him as an intelligent guerilla general but his true rôle was to liberate the country from Spanish rule, after which he should have left the task of government to others for he was incapable of organising the country. Alas! the Liberator became the Dictator. Yet the majority of the population was indifferent to any form of government as it had submitted to the superior groups of creoles and peninsulares. Also, as a staunch member of the *petite bourgeoisie*, Boussingault blamed the absence of a strong, governing middle class as was found in many European countries.[1] Without Bolivar, Boussingault had no further interest in South America but before leaving it was necessary to fulfil the wish of 1822 to exchange the dead volcanoes of his native Auvergne for the active volcanoes of the Andes. Thus started a long journey, a pilgrimage in fact, which began in December 1830 and did not end until February 1832.

The three parts of the Colombian Union were now held by dissident generals, Paez in Venezuela, Urdaneta in Colombia and Florez in Ecuador,[2] which meant that Boussingault was taking a risk in travelling in the uniform of Bolivar's former army. At the start of his journey roaming bands of former soldiery compelled him to await a more propitious moment before continuing his journey south to Quito, the capital of Ecuador. He thought it prudent to spend three months at Anserma and Cartago until the 24th March, 1831 when he set forth on horseback on a route through Cali, Popayan, Pasto and Tulcan. He was well prepared for the adventure for at some time he had climbed the volcano Tolime (5215 m), due west of Bogotá, which he had passed several times on his journeys from Bogotá to Cartago in the Magdalena valley.[3] His host at Popayan, Señor Varrela, possessed a collection of excellent quartz crystals which he believed were diamonds. Refusing to allow Boussingault to test them in a furnace he presented him with a kilogram of them which Boussingault threw away at the first river.[4] During the period March to December, 1831 he attempted the ascent of seven volcanoes, Purace (4700 m), Pasto (4266 m), Cumbal (4764 m) in Colombia, and those in Ecuador were Pinchincha (4697 m), Antisana (5704 m), Cotopaxi (5896 m) and Chimborazo (6310 m)[5] which was then considered to be the highest mountain in the world. The summit was reached on Cumbal, Pinchincha and Cotopaxi. Cumbal was partially active and to Boussingault "the scene was magical, flames emerging from a mass of snow, and a sky so dark blue that it was almost black".[6] There was an anticlimax to this wonderful experience when he discovered that the Indian carrier with the food had deserted. The two eggs which Boussingault was carrying were shared with two loyal Indians.

Under conditions which were often extremely uncomfortable, in which he was burnt on one side by puffs of hot vapour from fumaroles and frozen on the other by a glacial wind, he conducted experiments,[7] and the account of investigations on the slopes of Purace, which erupted in 1827, is typical of all the ascents. He presented a jar filled with snow to a jet of vapour from a fumarole of about thirty centimetres diameter and found that the condensed water did not contain a trace of hydrogen chloride although Gay-Lussac had found this gas in the vapours from Vesuvius.[8] This incident recalls the youthful adventure, with his friend Benoist of the mining school, at the burning coal mine of Ricamarie in 1819 when he was puzzled by the presence of ammonium chloride on the walls of a peasant's cottage. What was the source of hydrogen chloride, he asked, which was necessary to react with the ammonia from coal to form ammonium chloride? Gas from the vapour was also collected in a jar and shown to be mainly carbon dioxide with a little hydrogen sulphide and nitrogen. Unfortunately, Boussingault had not the opportunity to study fumaroles during an eruption, as did Gay-Lussac, when the composition of the gases was found to vary with temperature. Thus, it was found during the eruption of Etna in 1865 that the volcanic gases contained hydrogen chloride, sulphur dioxide, carbon dioxide, nitrogen and vaporised ammonium salts, but some of these gases disappeared as the temperature fell.[9] Also on Purace, Boussingault suspended a thermometer on an iron wire in the current of vapour issuing from a fumarole which registered the temperature of boiling water at that height. Under such discouraging conditions not only did he set up his barometer but also his portable laboratory, the contents of which, unfortunately, are not known, unlike those of some late eighteenth century mineralogists. As a geologist Boussingault was probably aware of the portable laboratory of his countryman, Guyton de Morveau (1737–1816) which was first used about 1780.[10] With his portable laboratory Boussingault was able to analyse minerals in the field such as a dolomite, normally a double carbonate of calcium and magnesium, in which magnesium carbonate had been replaced by manganese carbonate. At the mission station at Purace he set up his laboratory which did not surprise the priest who had witnessed a similar performance almost thirty years previously when Humboldt had been his guest.[11]

Boussingault entered Ecuador at Tulcan in July 1831 and soon established his headquarters at the capital, Quito. First a successful ascent of Pinchincha, 'the Vesuvius of Quito', but he was not so successful on Antisana which he had to abandon through snow-blindness.[12] Otherwise he suffered little inconvenience at these great heights because he was already acclimatised, and records how the young ladies danced incessantly throughout the night at balls at a height equal to that of the summit of Mont Blanc, without the slightest appearance of exhaustion.

Accompanying him on Pinchincha, and later Cotopaxi and Chimborazo, was a Colonel Hall who had served under Bolivar. After Bolivar's death Hall had establsihed a press in Venezuela in opposition to the Paez regime, but General Paez had him deported and Hall again set up his press, this time in Quito. A few months after the attempted ascent of Chimborazo with Boussingault, Hall was assassinated in a Quito street. Boussingault narrowly escaped with his life in October 1831 when a battalion of the army of General Flores in Quito protested about their pay. He was called in as a witness whilst the soldiers were being paid but dissatisfaction persisted and indiscriminate firing broke out. The subject of pay proved to be an excuse to demonstrate under the orders of a dissident general, Obando, who was stationed at Pasto, whither the demonstrating battalion retired.[13]

The Assault on Chimborazo[14]

Towards the end of 1831 Boussingault and Hall ascended Cotopaxi as a preliminary to attempting the ascent of Chimborazo the unconquered,

Fig. 9. Chimborazo, Ecuador, 6272 metres, from an engraving by Edward Whymper who was the first to reach the summit in 1880. (From his *Travels Amongst the Great Andes of the Equator* (1892, John Murray).)

regarded then as the highest mountain in the world. This adventure was to be the climax of Boussingault's ten years in South America. From Riobamba, the town from which all expeditions started to Chimborazo, including that of Humboldt and Bonpland, the two made for a small farm above the tree line at 3800 metres on December 14th, where they camped. On the following day they started up the gradual slope of the mountain and at a height of 4800 metres, the height of Mont Blanc, they covered their faces with taffeta masks to prevent snow blindness which afflicted Boussingault on Antisana. At 5115 metres they abandoned the attempt as the falling sonw was now more than a metre deep, and returned to their headquarters at the farm, but not before Boussingault had filled a bottle with snow to examine later the air in the pores of the snow which was found to contain only 17% of oxygen instead of the normal 21% He suggested, erroneously, that respiratory difficulties on a high mountain were perhaps due to the oxygen deficiency of air released from snow as it melted.

They then decided to ascend by the steep side, the route taken by Humboldt and Bonpland in 1802, although they were not aware that it was the same route. Not surprisingly, the guides who accompanied the climbers in 1802 were no longer to be found at the village of Calpi. During the ascent an unexpected difficulty arose when the pack-mules had to be left at about 5000 metres as they were too tired to proceed further, so, on foot, they reached what they believed to be 6004 metres, the greatest height then reached by man, but were prevented from proceeding further by the same crevasse that blocked the path of Humboldt and Bonpland. Whereas Boussingault was content to give barometer and thermometer readings en route, Humboldt, more geographically minded, had made a sketch map of the surrounding country as seen from the summit of a neighbouring mountain. Further, Humboldt could consult his thirty year-old notes to describe in great detail the route he had taken, for he was convinced that Boussingault and Hall had followed his path. By way of Calpi and San Juan, they had made for the Llanos de Sisgun, steppe-like country at the foot of Chimborazo. They had climbed the ridge, La Cuchilla, and proceeded along boulder-strewn terrain when they suddenly came to a crevasse only twenty metres wide but about four hundred metres deep, and again, this was the end of the ascent, after which they returned to Riobamba. Surrounded by formidable mountains, Riobamba should have presented as picturesque a scene as is found in Switzerland, but on the equator the sky was almost perpetually cloudy and Boussingault testified to the austere panorama from the town. Humboldt had experienced similar weather conditions which prevented him from making a trigonometrical measurement of the summit of Chimborazo which was continuously in cloud.

Boussingault's claim to have attained a height of 6004 metres was disputed by two authorities. Humboldt, in 1834, wrote to him from Berlin, whence he had been transferred by the Prussian court, suggesting that from Boussingault's barometer readings he had reached the same height as Humboldt which was not 6004 metres but 5878 metres.[15] It is interesting to note that both Humboldt and Boussingault read the mercury column of the barometer in 'lignes' (1 ligne = 2.55 mm) and their heights in 'toises' (1 toise = 1.949m) in the pre-Revolutionary units. The concept of a metric system was accepted by the National Assembly in 1791 and the necessary units were defined and adopted in 1799 but were not made compulsory until 1840; metrication was a slow process.

There was further controversy about fifty years later when Edward Whymper (1840–1911), English engraver and mountaineer, made the first successful ascent of Chimborazo in 1880 and found that there were two summits which had undoubtedly been obscured by cloud when previous expeditions had attempted the ascent. Whymper cast doubts on the heights reached not only by Boussingault but by Humboldt also, and considered that 5608 metres was

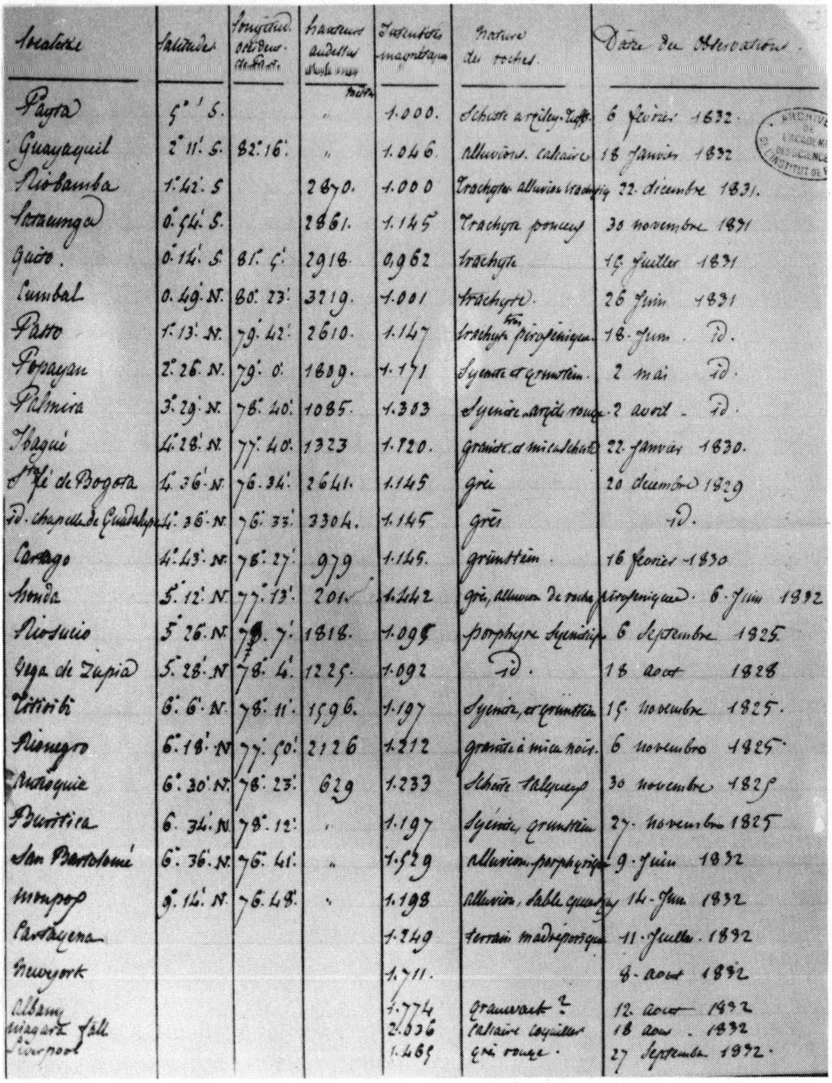

Localité	Latitude	Longitude occident. de Paris	Hauteur au dessus de la mer	Intensité magnétique	Nature des roches	Date des observations
Payta	5° 1' S.		"	1.000.	Schiste argileux talc	6 février 1832
Guayaquil	2° 11' S.	82° 16'	"	1.046	alluvions. calcaire	18 janvier 1832
Riobamba	1° 42' S		2870.	1.000	Trachyte. alluvion trachyt.	22. décembre 1831.
Latacunga	0° 54' S.		2861.	1.145	Trachyte ponceux	30 novembre 1831
Quito.	0° 14' S.	81° 5'	2918	0,962	trachyte	15 juillet 1831
Cumbal	0° 49' N.	80° 23'	3219.	1.001	trachyte	26 juin 1831
Pasto	1° 13' N.	79° 42'	2610.	1.147	trachyte pyroxénique	18 juin . id.
Popayan	2° 26' N.	79° 0'	1809.	1.171	Syenite et grunstein.	2 mai id.
Palmira	3° 29' N.	78° 40'	1085	1.303	Syenite. argile rouge	2 avril - id.
Ibagué	4° 28' N.	77° 40'	1323	1.120.	granite et micaschiste	22. janvier 1830.
Sta Fé de Bogota	4° 36' N.	76° 34'	2641.	1.145	grès	20 décembre 1829
id. chapelle de Guadalupe	4° 36' N.	76° 33'	3304.	1.145	grès	id.
Cartago	4° 43' N.	78° 27'	979	1.145.	grunstein.	16 février 1830
Honda	5° 12' N.	77° 13'	201.	1.242	grès, alluvion de roches pyroxéniques	6. juin 1832
Riosucio	5° 26' N.	78° 7'	1818.	1.098	porphyre syenitique	6 septembre 1825.
Vega de Supia	5° 28' N	78° 4'	1225.	1.092	id.	18 aout 1828
Titiribi	6° 6' N.	78° 11'	1596.	1.197	Syenite, et grunstein	15 novembre 1825.
Rionegro	6° 18' N	77° 50'	2126	1.212	granite à mica noir.	6 novembre 1825
Antioquia	6° 30' N.	78° 23'	629	1.233	schiste talqueux	30 novembre 1825
Buritica	6° 34' N	78° 12'	"	1.197	Syenite, grunstein	27. novembre 1825
San Bartolomé	6° 36' N.	76° 41'	"	1.529	alluvion porphyrique	9. juin 1832
Mompox	9° 14' N.	76° 48'	"	1.198	alluvion, sable quartzeux	14. juin 1832
Cartagena				1.249	terrain madréporique	11. juillet. 1832
Newyork				1.711.		8. aout 1832
Albany				1.774	Grauwacke ?	12. aout 1832
Niagara fall				2.036	calcaire coquiller	18 aout. 1832
Liverpool				1.485	grès rouge.	27 septembre 1832.

Fig. 10. Boussingault's records of latitude, longitude, height above sea level, strength of the earth's magnetic field and the nature of the rocks (Archives de l'Académie des Sciences, Paris).

the greatest height attained by them. Moreover, Whymper had found in the Botanical Department of the British Museum a specimen of a moss of the genus Orthotrichum collected by Hall at an alleged height of 5730 metres whereas Whymper noted that no mosses grew on Chimborazo above 5090 metres.[16] The evidence pointed to inaccurate barometer readings and interpretations, and subsequent measurements of heights have demonstrated that Boussingault's figures were too high. It was not until 1855 that the height of Everest (8845 m) was determined, yet in the Peruvian Andes there are many peaks higher than Chimborazo, the highest being Aconcagua (7010 m).

Boussingault continued his journey alone to the west coast port of Guayaquil from whence he embarked for the Peruvian port of Paita where be observed the guano beds along the edge of the bay. Many years later, as an agronomist, his thoughts returned to these guano beds on which he wrote an account with material supplied by his erstwhile colleague, Rivero,[17] who had become director of the national museum at the capital, Lima.

The Long Journey Back to Europe

Having achieved the objects of his journey, he was now anxious to leave South America and this anxiety is reflected in his notes which are barely informative of the places en route to the port of embarkation for Europe. He continued to measure the height, the latitude and longitude, the total magnetic field and to note the nature of the rocks, but towards the end of the journey we read only of dates, places and times of arrival and departure. From Paita he travelled northwards by sea to Buenaventura, calling on the way at Tumaco, then followed an overland route through familiar places, Cali, Cartago and Honda where he boarded a river boat and sailed down the rio Magdalena to Cartagena where, on July 11th, 1832 he embarked on the 'Medina', arriving in New York on August 7th. Before leaving for Liverpool on the 'Ajax' he visited the Niagara Falls and eventually arrived in Paris about the first week in October. The South American phase of his life was over but it was to influence him for the rest of his days.[18]

An interesting and revealing comparison can be made between the itineraries of Boussingault and Humboldt in South America although it must be recognised that Humboldt, as ever, was completely independent, financially secure and socially uncommitted, for in spite of his undoubted attraction to women, he never married. Appointments to the Colombian government and the Colombian Mining Company restricted Boussingault's independence, and the exploratory aspect of his South American period could not produce the results achieved by Humboldt; there was no event so outstanding as Humboldt's confirmation of the presence of the Casiquiere canal, in May 1800, linking the river basins of the Amazon and the Orinoco during which his boat was carried overland from one tributary to another. But there was some correspondence in the journeys thereafter, the expedition to the volcanoes of Ecuador and the visit to the Peru coast. Even the Meta expedition had some point of contact when Boussingault measured the latitude and longitude of its confluence with the Orinoco which Humboldt was unable to accomplish because of the presence of heavy cloud. The return to the United States of America was, however, a significant contrast. Whereas Humboldt had been enthusiastically welcomed in Philadelphia and was even received by the President, Boussingault's arrival appears to have been unheralded and he was content with a private visit to Niagara. Yet his reputation in South America gradually grew as his memoirs of the country, its people and its products appeared in scientific journals for another fifty years. Both Boussingault and Humboldt were from countries whose representatives were welcomed by the rebellious Spanish colonies. Bolivar preferred to recruit French officers in his forces, especially those who had emigrated after serving at Waterloo. It is highly probable that Humboldt would have returned, if not to South America certainly to Mexico, if Boussingault had decided to return also. His failure to return may have induced Humboldt to persuade a number of German naturalists to visit the sub-continent, many of whom settled there permanently and founded museums. Lack of colonies may have been responsible for the German tradition in South America.

What of the others who set forth from Antwerp with Boussingault in 1822? No more is heard of the entomologist Bourdon, apart from the scurrilous episode when Boussingault was prostrate with malaria after the Meta expedition. Goudot, the botanist, greatly admired the indigenous Indian population and it is not surprising that he did not return to Europe but remained in Colombia and carried out extensive field trials on the cultivation of the parsnip-like vegetable, the arracacha, a member of the family Umbelliferae which includes carrots, parsnips, celery as well as a host of wild plants such as cow-parsley and fennel. Following the potato disease of 1845, which was catastrophic in Ireland, Goudot suggested the substitution of the arracacha, but field trials demonstrated that the plant required 10–14 months to mature and success depended on constant periods of sunlight which was possible in the tropics.[19] Rivero left Bogotá in 1825 for his home country, Peru, where he became director-general of mines and then director of the national museum at Lima. When Boussingault was leaving for Europe, Rivero became a member of the Peruvian Congress, his new political career culminating in his appointment as consul-general in Belgium.

Apart from Boussingault the most outstanding member of the team was the doctor-artist-naturalist François Roulin who left South America in 1828 after discovering two new species of sting-ray and a new species of tapir, Roulin's tapir, the Andes tapir or the hairy tapir. In 1829 he wrote a series of scientific articles for *le Temps* and *le Globe* which were carried in these newspapers for several years; they were summaries of the meetings of the Académie des Sciences which Roulin developed into a journal in 1835, *Comptes rendus des séances hebdomadaires de l'Académie des Sciences* which he edited until 1862 when he became librarian of the Institut de France.

The Rôle of Humboldt

To gauge the European reaction to the expedition several aspects of the expedition's organisation should be considered. First of all there is the rôle of Humboldt either as 'non-playing captain' or as 'public relations officer'. Perhaps P.R.O. is a more suitable description of his function because the expedition lacked a coordinator. There was little or no cooperation between the members; exceptions, of course, can be cited such as the joint attempt to elucidate the nature of curanine and the examination of Gay-Lussite. On the other hand what function did Goudot perform when he accompanied Boussingault to the Choco? We hear of no botanical collection having been made. They were members of a group which needed a programme coordinator who would, through occasional meetings and correspondence, assemble a picture of Colombia and its scientific importance and problems. Humboldt could have assumed this rôle and, indeed, was anxious to return to the New World, and we can only conjecture that he was engrossed in the production of his masterpiece *Cosmos* a wonderfully illustrated description of the world in five volumes. As a public relations officer he was superb. Not only did he correspond with the team but sent them the latest scientific journals. More than that though, he ensured that Boussingault's scientific contributions were published and wrote memoirs to the Académie des Sciences reminding them of the various projects which were being undertaken in Colombia. Sensitive to the wishes of others he conveyed news of the party to relatives and visited them to hear more intimate details from Boussingault's letters to his brother-in-law, Vaudet and sister Jeannette. He arranged for the host of travellers known to him to convey letters to and from South America; briefly he acted as a post office. The more we read about Humboldt the greater our appreciation of his fine qualities not only as a scientist but as a cultured personality possessed of the finer feelings for humanity in general and for his friends and acquaintances in particular. His letters bear testimony to such qualities.[20]

On returning to Europe Humboldt repeatedly encouraged Boussingault to publish his memoirs on ten years' sojourn in South America but without success.[21] It followed that Boussingault had sufficient material to occupy the remainder of his life so that articles on South America appeared regularly in scientific journals. Towards the end of his life he dictated his memoirs to a married daughter but these went no further than 1832, and at that distance of time they could hardly be regarded as scientific contributions but a highly interesting and entertaining travelogue.[22]

An Assessment of Boussingault's Work in South America

The extent of his researches during these ten years is remarkable considering the scanty formal education which he had received for a scientific vocation, limited to the facilities of a mining school. He approached South America with no definite field of research unless South America was his field of research which it appeared to be. There was so much to be investigated over too wide a field that during those ten years he developed the polymath tendency which was so characteristic throughout his life. A perusal of the titles of the forty-five scientific papers written between 1822 and 1832 confirm this impression as they covered geology, mining, mineralogy, chemistry, medicine, meteorology and volcanology. Such a programme might not be conducive to work of permanent value and it is, therefore, a salutary thought that some of his work is cited today as being either of a pioneer nature, or of value in elucidating and explaining previous discoveries or processes. In the latter class can be included his investigaition of the Patio process for the extraction of silver from its ores. The silver ore, usually a sulphide, was first crushed to a fine powder and rendered a muddy mass by the addition of water. The ore was then carted to a patio which had been made impermeable to mercury by the use of asphalt and cement when constructing the floor. The round flat cake of ore, sometimes fifteen metres across and known as a 'torta' was treated with salt followed by 'magistral' or anhydrous copper sulphate. Mercury was then added to the torta and trodden in by as many as twenty-five mules to ensure an intimate mixture. For the mules this was fatiguing work and they were rested at intervals during the day during which the torta was turned over by workers using shovels. After three to six weeks of mixing, depending on the atmospheric temperature, the contents of the torta were washed with water in vats where the resulting silver amalgam collected at the bottom; it was then scooped out and heated in iron retorts to expel the mercury and so obtain the silver. It was a wasteful process as for every unit of silver recovered one and a half units of mercury were lost.[23]

The chemical reactions taking place in the torta were controversial as the intermediate compounds suggested by various workers could not be isolated or identified. In a complex series of operations at least one problem was solved, that is whether the operative agent was cuprous chloride or, as Boussingault found, cupric chloride, produced by the interaction of copper sulphate and salt. Briefly, he found that cupric chloride reacted with silver sulphide in the ore to form silver chloride which was reduced by mercury to silver.

Ten years work over so wide a field was bound to produce some erroneous hypotheses such as the method of finding the mean annual temperature of a place in the tropics, This was to find the temperature of the layer of soil at a depth of one foot below the surface.[24] His statistics were most convincing as his results could be compared with the abundance of temperature readings taken over the years at these places. But his apparent success depended on the slight variability of temperature at the tropics so that all temperature readings tended to some mean value but this value was not, as believed, a constant figure, e.g. 27 degrees or 28 degrees. A fault of Boussingault's method was that be omitted his reasons for explaining a result – what was his 'thinking' on the method used? How did he conclude that the mean temperature at a station was the temperature of a layer one foot below the surface? However, this paper and another about the Patio process gained him a doctorate at the *Sorbonne*, in Paris in 1834.

If the stages of development of a science can be expressed as description, generalisation and hypothesis, then much of Boussingault's work in South America must be included in the category of description, which is not surprising. However well-known South America was to the Spanish people, the ban on foreigners had long rendered it a land of mystery so that additions to the discoveries of Humboldt and Bonpland were eagerly sought and much appreciated by the *Académie des Sciences*. Thanks to the patronage extended to him by Humboldt all his communications were published. The sub-continent was at the descriptive stage and more information on various subjects was needed before valid generalisations could be made. Also, Boussingault's level of scientific attainment at the time was more related to the descriptive stage; this particularly applies to his numerous analyses of rocks, ores and water. An example of an erroneous hypothesis occurring in two of his articles was that specimens of native gold, found in Colombia, contained compounds of gold and silver when, in fact, they were alloys and therefore mixtures of the two metals and not compounds. When copper is burnt it first forms brown cuprous oxide and finally black cupric oxide, but between the two extremes are various colour changes due to a mixture of these two oxides. In 1809 there arose a controversy between Claude Louis Berthollet, who claimed that these intermediate colour changes signified the presence of intermediate compounds. His contemporary, Joseph Louis Proust, correctly reasoned that these colour changes were due to varying mixtures of the two oxides. Likewise, Boussingault had mistaken mixtures of gold and silver for compounds because of their simple proportions, but in his articles he made no mention of what is now known as the Proust-Berthollet controversy.[25]

Scientifically, the period 1822–1832 were his formative years and South America was his university, but it was a university in which positivist ideas were implicit. It was Auguste Comte (1798–1857) who introduced a scientific philosophy which can be described as positivistic in that we should restrict ourselves to what we can ascertain positively and not speculate about the nature of the world beyond the facts of our own observation. Our sense experiences tell us all that we can possibly know positively – one set of observations will explain another set, but our explanations should not go beyond the observational framework. The proper rôle of a scientist is to state his evidence but not to generalise or theorise about them. Briefly, such were the tenets of positivism and although they were not presented in book form until Comte published his *Cours de philosophe positive* from 1830 to 1842, positivistic ideas had been accepted by a number of French scientists in spite of the fact that they allowed no place for generalisation and prediction. As far as can be ascertained Boussingault did not mention Comte or positivism in his writings but his interpretation, or lack of it, of his own experiments suggests that he had been influenced by Comte and his school. This, then was the mental background which he carried back to Europe in 1832 and with which he made his début as a pioneer agricultural chemist after a brief and unhappy half-year at Lyon. Like Humboldt, he had trained as a scientist at a mining school, and like Humboldt, who made a life study of geophysics for which he had never been trained, so Boussingault devoted much of his long life to agricultural science for which he, also, had never been trained.

CHAPTER 5

PARIS, LYON AND PARIS AGAIN

The traveller returning to his mother country, after a long sojourn overseas, finds it necessary to make mental adjustments in integrating into the older conformist society from which he sprang. He makes comparisons, sometimes unfavourable, with the freedom and limitless horizons, the sociability and opportunities, and, like many today, Boussingault was no exception. France was an anticlimax after the adventurous years and not only was he irritated by the conventions of the metropolis but he was astounded by the intrigues associated with appointments to scientific posts, The *Muséum d'histoire naturelle* was singled out as a particular example from which it is concluded that he applied unsuccessfully for a post there. More specific was his attempt to canvass the *Institut de France* on behalf of his old Alsace geologist friend, Phillipe Voltz, who had, in 1822, initiated the chain of recommendations which enabled Boussingault to become part of the South American expedition in that year. Voltz had applied for the chair of geology at the University of Strasbourg and Boussingault's recommendation had been received sympathetically and encouragingly by the dean of the faculty of Sciences at the *Sorbonne*, no other than Louis Thenard. Imagine the shock when he discovered that another candidate, Augustus Daubrée, had been appointed [1] – although it can be said that Daubrée, a younger man, filled the post with credit and honour. Boussingault clearly unburdened himself to Humboldt who wisely reminded him that annoyances in South America and France were very much the same but that they were more bearable in the solitude of forests and mountains. It was also a return to his family as he lodged at the residential complex built by his brother-in-law, Sylvestre Vaudet, at No. 1 rue du Parc-Royal in the Marais district of Paris. Boussingault's parents were also living there, having sold the tobacconist business and the house in the rue de la Parcheminerie in 1822.

Hardly had he returned to France when over-anxiety and over-sensitivity on Humboldt's part produced a temporary breach between him and Boussingault. In retrospect it could be predicted that such rifts were unavoidable whilst one was in Paris and the other in Potsdam, rifts which could be avoided by personal contact. Humboldt was hoping for a speedy return of his friend to South America but not before he had written the memoirs of his ten years there. Yet Boussingault had never recovered from Humboldt's attempts to be friendly and appeared overawed by him to the chagrin of Humboldt who pleaded with him to write and assure him that he harboured no grudge against him, to treat him somewhat as a friend, and to avoid 'Dear Sir' which he only used when writing to the English![2]

Boussingault had developed a slight paranoia, due perhaps to ten years of comparative isolation, or to the ease with which his writings had attained publicity in the scientific world of Paris, thanks to Humboldt. Whatever the reason, Vaudet had cause to write that stay-at-homes bowed down before the celebrated traveller, M. Boussingault, because they were unable to see such nice things as the poisoned arrows of savages and the snakes.[3] There was a finer sarcasm from Humboldt when asking Boussingault to convey his regards to MM. Arago, Gay-Lussac, Dulong, Poisson and if he had been introduced, to M. Boussingault, the most learned traveller of our time, filled with talent and fearlessness![4] Humboldt intervened to stop Boussingault from making an early and premature application for membership of the *Institut de Frence.*[5] The *Institut* was founded in 1795 as a means of resuscitating the various cultural academies which had closed during the Revolution, some of them dating from the seventeenth century. It consists of five academies encouraging the study of the French language, science, fine-arts, literature, and moral and political science. Membership is by election and it is keenly sought as it is the hall-mark of an outstanding practitioner. Whenever Boussingault appended his signature to an official document he rarely described himself in professional or political terms such as 'professor', or 'deputy' or '*Conseiller d'Etat*', but usually as '*membre de l'Institut*. Among the various sections of the *Acadêmie des Sciences*, Agriculture was recognised as a separate section – *Economie rurale*. Having carried out agricultural trials on the Alsace farm of the LeBels, Boussingault applied for election in 1838 when a vacancy occurred owing to the death of the botanist Henri Tessier. Not only was he not elected but he did not receive a single vote, the successful candidate being Jean Audouin who attended the farewell dinner in 1822 at chez Véry. Humboldt, alarmed by this result, spoke to the physicist Jean Baptiste Biot, an influential member of the *Institut* who replied that he did not know M. Boussingault and would be glad to be introduced so that he could speak to him about his excellent observations of temperature and pressure. Humboldt reported the conversation in a letter to Boussingault with a warning that he was to meet M. Biot on the morrow but there was no need to be disagreeable to him.[7] Humboldt's salutary lesson was timely, for the following year saw Boussingault elected easily as a result of a vacancy caused by the death of the veterinarian Jean Baptiste Huzard.[8]

A further difference occurred when Humboldt wrote that he had few regrets on leaving America because Boussingault had travelled in the country for the past ten years, and that he would now report on magnetic and meteorological phenomena, geology, altitude, the soil and the chemical nature of New World products.[9] These remarks were contained in the preface of a book about the development of nautical astronomy in the fifteenth and sixteenth centuries. It was published in Paris between 1836 and 1839, with the title *Examen critique de l'histoire de la géographie du nouveau continent et des progrès de l'astronomie nautique aux quinzième et seizième siècles*. Even Humboldt realised that Boussingault was being gently informed on how to conduct his scientific life, so when Boussingault asked permission to dedicate a proposed volume on volcanoes to Humboldt, the latter declined as he felt that there would be objections to the passage in the preface of his *Examen critique* ... and indeed, Boussingault was displeased.[10] Yet he continued to receive encouragement from Humboldt and sometimes criticism of his methods and of the accuracy of his results.

Again, the distance between them may have contributed to Boussingault's failure to obtain a suitable post in Paris which was then the intellectual capital not only of France but of the whole of Europe. That most enlightened of Ministers of Public Instruction, François Guizot, received a letter from Humboldt recommending Boussingault for a post but to no avail. The reasons we can only surmise; was he temperamentally unsuited? Was Humboldt's patronage a handicap, especially now that he had returned to Prussia? Had there been a certain dilettantism in his career? Was he a geologist, a meteorologist, a chemist or an agronomist since he had contributed to journals on all these subjects? It appears that at the suggestion of the chemist Louis Jacques Thenard he accepted the chair of chemistry at the newly resuscitated faculty of Sciences at the *Académie de Lyon*. Humboldt was displeased and probably regarded it as a decline in his professional influence for he wrote, "I have not the courage to advise you to break with Thenard and Lyon". This long letter of good counsel begins by thanking Boussingault for his "excellent and instructive letter of 23rd August which consists of two lines"![11]

The canalising of his energies in the direction of agricultural science was not to take effect for another year, but meanwhile there occurred an event which fixed his scientific course for the best part of his life. On January 7th, 1835 he married Adéle LeBel at the Protestant church of Kuntzenhausen in Alsace.[12] He had known her since his Lobsann days when he was a frequent

Fig. 11. Adèle Boussingault (née LeBel). (Collection: Mme F. Boussingault.)

visitor to her father's farm Bechelbronn which was in the commune of Lampertsloch of which her father was mayor. For Boussingault the marriage meant that all thoughts of returning to South America were now abandoned, and there would never be a meeting with Humboldt in Mexico, who received the news of the marriage with his usual calm, congratulating Boussingault but hoping that marriage would not detract from his scientific work.[13] After the wedding Boussingault and his Adèle went immediately to Lyon.[14]

The Académie de Lyon

The University of Lyon was not formally established until 1896 [15] although various schools of university standing existed before the Revolution. The Municipality of Lyon demanded the reopening of the schools of Law and Medicine, but instead was given in 1809 an *Académie de Lyon* with faculties of Arts, Sciences and Theology.[16] The students were poorly taught and housed and few received degrees so that at the end of 1815 government approval was obtained for closing the faculties of Arts and Sciences, there being no protest against the closure until 1834,[17] so apathetic did the Lyonnais appear. No reason is given for this strange and unnatural indifference but certain tentative conclusions may be drawn; firstly, that it was the result of the severe economic depression which France underwent for about twenty years after the end of the Napoleonic wars. This applied especially to the silk weavers of Lyon who, having won a wage agreement with the merchants, found that it was repudiated. Popular discontent was repressed in 1831 with loss of life when a large body of troops was dispatched to Lyon, and this was followed in February 1834 by further rioting which was crushed by ten thousand troops.[18] In periods of stress and violence thoughts of cultural progress tend to fade into the background. A second, and perhaps related reason for this apathy was a conspiracy of silence against the faculty of Sciences as shown by the neglect on the part of local historians to record adequately the development and vicissitudes of the faculty. Three works on the history of the University, published as early as 1842 and as late as 1922, gloss over the existence of the faculty of Sciences in a line or two.[19] Jean Joseph Soulacroix was Rector of the *Académie de Lyon* and also headmaster of the Collège Royal de Lyon. At a College prize giving in 1834 he announced the forthcoming opening of the new faculty and although he was personally enthusiastic about the project he expressed himself lukewarmly, warning his audience that Science was not sufficient and that the teacher should remind his young disciples of the duties which fit them for a place in society.[20] Obviously, his enthusiasm for the new faculty was not to be advertised in Lyon. The source of this conspiracy of silence may be inferred from the reaction to his address in 1838 on the reopening of the faculty of Arts. He is reported to have said that youth must be attracted not by the voice of the cloister but by the voice of the world. Youth, he said, should be shown examples of great men whom they admire and hope to emulate, rather than examples of the heavenly beatitudes. For this vigorous defence of the practical world Soulacroix was attacked by the Church, accused of impiety and even of atheism.[21]

In applying for the re-establishment of the Sciences faculty the mayor of Lyon, Dr Prunelle, had the valued support of the Préfet of the Rhône département, Adrian Gasparin, a remarkable man who had retired from the army following a serious wound at the battle of Eylau in 1807 and had become deputy, préfet, and later a peer, Minister of the Interior and finally Director of the *Institut national agronomique* at Versailles. Gasparin's plea was:

> All intelligent men have for a long time protested against intellectual centralisation no less than against administrative centralisation. In France the scientific spirit has been cast in a single mould, and if it has made remarkable progress it has made it in one direction only. It lacks the variety and competitive spirit so characteristic of the German scholar. The time has come to break with this monotony. But we can only do this by putting up schools in a town which by its importance has great influence and power, rivalling to a certain degree that of the capital.[22]

The mayor's request was granted by an Act of the king Louis-Philippe of 9th December, 1833 but owing to the lack of government funds the faculty could not start until January 1835.[23] Seven professorial chairs were to be established and in proposing names of professors he included Achille Penot of Mulhouse "for general chemistry and physics". Penot was known for new methods of chemical analysis including the use of potassium iodide – starch paper as an external indicator.

However, a Ministerial notice of July 21st, 1834 named the professors among whom Boussingault appeared as professor of chemistry.[24] For a faculty of Sciences opening under such inauspicious circumstances the calibre of the professors was most encouraging, with Cournot (mathematics and mechanics) afterwards Rector of Grenoble, and then Dijon; Tabereau (physics) who was assistant secretary to the *Académie des Sciences, Belles-Lettres et Arts de Lyon,* a thriving institution then as today; Jourdan (zoology), director of the Muséum d'histoire naturelle at Lyon; Seringe (botany), director of the local botanical garden, now only a grassed open space at the end of the rue du Jardin des Plantes, the present botanical garden being situated on the opposite bank of the Rhône; Fournet (mineralogy and geology) who was the public assayer at the Lyon mint; and Clerc (astronomy) who had been a professor in the faculty when it had closed in 1815. In addition to the chair of chemistry Boussingault was appointed dean of the faculty.[25] Thus was the stage set for his arrival in January 1835 but it was not the Lyon of today with its majestic streets, squares and buildings, which were the result of extensive replanning in 1855 by Claude Vaisse who did for Lyon what 'baron' Georges Haussmann achieved in Paris by constructing the boulevards.

The new faculty was inaugurated at the Palais St Pierre on January 29th, 1835 in the presence of s brilliant gathering[26] although it was reported that many local scholars were absent. The Palais St Pierre, a former seventeenth-century monastery, housed the art gallery and the museum of natural history and pending the erection of faculty buildings many of the courses were to be held there. The Rector, Soulacroix, opened the ceremony with an enthusiastic address which prematurely overstressed the benefits likely to accrue to local industry now that the faculty of Sciences had been re-established. Boussingault followed and suitably modified this enthusiasm by reminding the audience that although the tanning industry had seen the advantages of chemical research, the dyeing industry had yet to witness such improvements. Boussingault's address revealed his attitude at the time towards science and scientists which can be summed up by the terms 'dedication' and 'nationalism', for he emphasised the need for young people to dedicate themselves to science and cited examples of scientists who had perished in the pursuit of their investigations.

He also stressed the rôle of chemistry during the Napoleonic wars when a sea blockade prevented the import of the seaweed barilla which was collected on the Mediterranean coast of Spain and sent to Marseilles to be used in the production of soap. By the LeBlanc process, first announced in 1789, France was able to dispense with barilla by producing from common salt the soda necessary for soap manufacture. The use of beet sugar in place of overseas cane sugar, and the search for more and better saltpetre for gunpowder were among other of his examples of how a nation, under duress, could be made self-sufficient.

If the address provoked the suspicion that Boussingault regarded himself as the embodiment of Science, it was confirmed by his choice of James Watt as an example of a scientist. Watt, he reminded his audience, was not a physicist but an instrument maker. Success in the arts, he continued, was reserved for the manufacturer and the skilled artisan, for the rôle of the savant was to teach the young who were destined for an industrial career. With this blundering remark he almost sabotaged the faculty. Did he not know of the existence and the work of the *Ecole de la Martinière* in Lyon? His audience was certainly aware of it. It will be recalled that the faculty of Sciences was closed in 1815 and it was earlier suggested that local apathy was responsible for its continued closure. Perhaps there was another reason. The lacuna caused by the 1815 closure was soon filled in a manner more in keeping, perhaps, with the aspirations of the young Lyonnais. From 1816 onwards various trade and technical schools were established of which the most successful was the '*Martinière*', one of the outstanding features of which was the employment of *répétiteurs* or tutors who supervised the study hours and fully reviewed the lessons of each day, as was the practice at the more famous *Ecole polytechnique* in Paris. Former students were to be found as shop managers and foremen in Lyon, and others held senior positions in commerce, industry and banking throughout France. What was the advantage of pursuing a university course to become a savant who would assist '*Martinière*' students to enter more lucrative careers?[27] Such a thought may have registered in the minds of the students and audience for the reporter of the Lyon newspaper, *Le Réparateur*, to have written that he noted the lack of interest on the faces of the *Académie* students and concluded, "As for me, who had never before listened to an academic address, I left the Palais without any desire to become a member of the Académie"! However, the occasion was greeted with enthusiasm by the pro-government newspaper, *Le Courrier de Lyon*, which continued to give support to the faculty with news items and editorials.

Following the opening of the faculty, the story of Boussingault at Lyon is briefly as follows. His chemistry course was intended to occupy the second term of the year but his responsibilities as dean obliged him to be at Lyon throughout the year. From January to June 1835 he was preparing his course and by the end of February his chemical reagents were ready.[28] His course opened on June 11th and closed at the end of July, having consisted of thirteen lectures on Mondays and Fridays at 11.00 a.m.[29] He then left Lyon and was never seen there again, yet he failed to send in a letter of resignation until another fifteen months had elapsed.[30]

Soon after their arrival in Lyon, Boussingault and the professor of physics, Tabereau, began protesting about their lecture and laboratory accommodation at the Palais St Pierre. The Rector had written to the Minister as far back as the previous March outlining the necessary accommodation,[31] two lecture rooms and a laboratory each for physics and chemistry. But Prunelle, the mayor, had informed the Minister that adequate laboratory space was available at the Palais[32] and the Minister used this information in reminding the Rector that the accommodation was considered suitable.[33] The Rector promptly replied that Prunelle was mistaken and quoted Boussingault as saying that there was nothing with which to carry out the simplest experiment and that Tabereau had no room to assemble the physics apparatus once it had been unpacked. The Rector insisted that these subjects could not commence under such conditions and, in addition, the funds allocated to the faculty were insufficient and thus a supplementary allocation was requested.[34] This was refused by the Minister[35] but Soulacroix pressed forward with the erection of a new building for physics and chemistry, the opening date of which was postponed several times until November 1835.[36] It was erected adjacent to the *Collège Royal,*[37] now the *lycée Ampère*, on the quai de Retz, today known as the quai Jean Moulin. An elderly teacher of the *lycée Ampère* recollected in 1971 that the seats of the lecture theatres were removed in 1914, "just before World War I".

Boussingault Leaves Lyon

Meanwhile, Boussingault had taken up residence with his wife at No. 8 rue Mézières, near the Jardin du Luxembourg in Paris where he established a laboratory. The Minister had granted him an extension of leave, probably because their first child was expected.[38] Tabereau was to teach both physics and chemistry until Boussingault's return but he fell ill the day before term opened and so there were no courses in these subjects.[39] This did not pass unnoticed in the city for there was a letter on the crisis in *Le Courrier de Lyon.*[40] Tabereau recovered sufficiently to start the physics course on February 2nd, whilst the chemistry lessons were undertaken by Faurie, the new professor of mathematics who had replaced Cournot, now Rector of Grenoble. To add to Soulacroix' difficulties, the professor of botany, Seringe, proposed a rearrangement of his programme by which he would teach the nine months' course in the month of February but this was rejected by the Minister.[41] Seringe had the excellent precedent of Augustin Pyrame de Candolle of Montpellier who taught six days of the week with field work on Sundays so that he completed his course in two months, after which he was free to conduct his famous surveys of the botany of France.[42]

A sad event caused a further postponement of Boussingault's return to Lyon, for the daughter, born in January, had died soon after birth. Soulacroix was undoubtedly amazed to read in the same letter from Boussingault that his erring professor of chemistry had been made a member of the Legion of Honour![43] A week later he wrote suggesting that he might resign but did not tender his resignation on the grounds that he could not find a suitable successor in Paris and that Lyon applicants were unsuitable.[44] This disparagement of local talent was repeated in a further letter[45] but he was completely unjustified in denigrating the scientific community of Lyon who fostered the cultural life of the city by their work for the *Académie des Sciences, Belles-Lettres et Arts de Lyon*, of which Boussingault was never a member. At last, the Minister, who had acted as an intermediary between Boussingault and his

Rector, decided that he had overstayed his leave [46] and that he should resign, but Boussingault made a desperate effort to retrieve the situation by suggesting that his course should extend over two terms, the second term to be devoted to organic chemistry. The appointment of an assistant professor was also mentioned and it was asked why a Lyon professor received one-quarter of the salary of a Paris assistant professor who gave only half his number of lectures. In the same letter, he recommended Henri Victor Regnault as his successor and, scathing on the subject of Lyon salaries, suggested that Regnault might obtain extra employment locally as a mining engineer, as part of his education had been pursued at the *Ecole des Mines*, Paris.[47] Leaving the Giessen laboratory of Justus von Liebig, Regnault taught the chemistry course of 1836 [48] but from the beginning of 1837 a permanent appointment was made in the person of Armand Bineau who had been a demonstrator at the *Collège de France.*[49]

It may not have been a ruse to bring Boussingault back to Lyon, but his name appeared on a list of 36 jurors for the second assize of 1836.[50] He had previously received a summons to appear on a jury [51] and had appealed, apparently successfully, against attendance. At last, in October 1836, he sent a letter of resignation to the Rector after which Tabereau was appointed dean of the Faculty, taking over from the aged and faithful Clerc who had acted as dean during most of 1836.[52]

There remains a question which has not been satisfactorily answered: why did Boussingault leave Lyon? In spite of the low professional status of the *Académie de Lyon* he was not only a professor but the dean of the faculty and in a year or two the post might have led to advancement and an appointment in Paris. Although he eschewed the social life of Lyon he once wrote to the Rector, perhaps with lack of sincerity, that he preferred Lyon to Paris.[53] The "friendly and Germanic Mme Boussingault', Humboldt's phrese.[54] may have preferred cosmopolitan Paris to French-speaking Lyon. The low salary might have been compensated by other employment according to the prevailing '*cumul*' system by which university staff accepted several posts, as in the case of Tabereau who also had an appointment at the '*Martinière*'. It was possible that Boussingault realised that the Lyon appointment was an error, and even Humboldt thought so, because he could see no bright future there, suffering as it did from the apathy of the city and the antipathy of the Church. Yet his introvert nature would not allow him to rationalise the impasse and hand in his resignation in 1835, and so the whole sorry business dragged out to the end of 1836 with no reputation enhanced save that of the Rector, Soulacroix.

The lack of university autonomy was amazing for almost every detail of administration had to be referred to the Minister of Publuc Instruction, and Guizot was a genius for detail, a tradition started by Napoleon in 1808. Yet in the eighteen-thirties many in England wondered how long they would have to wait before a similar scholar-statesman would guide and develop the public education of their land.[55] Apart from his statesmanship, Guizot was an historian who wrote not only on French and European history but also on English history, the result of his exile in London after the 1848 revolution.

Boussingault's brief sojourn in Lyon could not have been conducive to research although he continued his work on the supposed presence of a hydrocarbon in the atmosphere related to a malarial miasma.[56] There were analyses of Rhône water and that of the Royes spring which eventually flowed into the Rhône. They arose from a proposal of the Lyon Municipality to introduce a domestic water supply in addition to the existing fountains which were supplied by private companies from the Royes spring. As the Municipality intended obtaining the extra water from the Rhône, the companies feared the loss of a contract and one of them, Loubayon, Vetter et Cie, engaged Boussingault to make a comparative analysis of Rhône and Royes water. He was unnecessarily guarded in his conclusions for he declared that Royes water was as good as Rhône water,[57] a claim that must be seen against the background of local belief that Rhône water was pure, a belief to which he replied that only distilled water was pure. Six years later Boussingault published his results in his *Economie rurale*[58] and one cannot understand his reticence in not revealing the true facts in 1836 that the Royes spring contained two and a half times the weight of solid matter as that found in the Rhône, and further, that whereas the Rhône water contained no nitrates Royes water contained a trace, due perhaps to farm sewage. It is difficult at this distance of time to appreciate the anxieties and pressures which he experienced, but there is little doubt that Lyon was the most unsatisfactory period of his career and that, on the whole, he acted in a most regrettable manner.

Boussingault's motives in leaving Lyon are further confused by his attempts to establish two separate but eventually related careers in Paris and Alsace. It is remarkable that whilst officially the dean and professor of chemistry at Lyon he had already started what was to prove a series of fundamental field trials at his father-in-law's farm, Bechelbronn, from 1836 to 1848 in the realms of crop rotation and the origin of fat in herbivores. From his scientific papers we may infer that the remainder of 1835 was spent in visits to Bechelbronn and in the writing of papers on South American topics such as cocoa, the banana, and tropical rain, but the most provocative was a paper comparing the manner in which the same food crops grew at the equator and in the temperate zone.[59] In practice he was comparing the time taken by crops to mature at Bogotá and Bechelbronn. He found an interesting relationship between two clearly defined quantities, first, the number of days between germination and maturity, and secondly, the mean temperature during this period. The product of these values was found to be a constant for the same food crop; for wheat at Bechelbronn, Bogotá, Cincinatti and New York State the product was 1700; for barley it was 1700 also; for maize 2500 and for potatoes it was 3000. Boussingault admitted that his observations were not sufficiently numerous but wished to draw attention to an interesting relationship, even if it did not justify the term 'hypothesis'. It was merely one aspect of a general rule for all types of vegetation but to his contemporaries these results elevated the relationship to a natural law.[60] He appears never to have returned to this topic or even held to the relationship seriously, for growth is dependent not only on temperature but on light, the nature of the soil, and moisture.

Boussingault at the Sorbonne and Collaboration with J.B. Dumas

At the *Sorbonne* it was Thenard's custom to teach the first half-year of the chemistry course, leaving the second half to his assistant professor, Jean Baptiste Dumas,[61] but in October 1836 Thenard announced that he would not be teaching the first half-year of the course in 1837, the normal procedure of a professor of fifty-nine years who having acquired several posts under the '*cumul*' system was now contemplating a reduction in his teaching load. Therefore Dumas taught the first half of the course in 1837 but the lecturer for the second half could not be confirmed because of Thenard's absence from a meeting of the faculty of Sciences, and thus Dumas' replacement for the second half of the course could not be advertised on the college posters. But on March 5th, 1837 Thenard wrote to the Minister recommending the appointment of Boussingault to replace Dumas,[62] an appointment which was confirmed by the Minister by letter of March 31st. The *Sorbonne* appointment was the beginning of a scientific collaboration with Dumas in the diverse realms of atmospheric composition, the chemical composition of plant and animal materials, and the production of fat in farm animals. If, formerly, Boussingault was known at all, it was through his co-operation with Dumas in the quantitative determination of the composition of the atmosphere in 1840,[63] an experiment which has been reproduced in various textbooks of chemistry. Chemists then held three opinions on the composition and nature of atmospheric air, some believing that it was a compound of oxygen and nitrogen as suggested earlier by Sir Humphrey Davy; others, such as Liebig and Gay-Lussac regarded it as a mixture of constant composition; whilst John Dalton thought it was a variable mixture, the proportion of oxygen decreasing with altitude, although Gay-Lussac in 1804 in a balloon ascent had shown that the proportions of oxygen and nitrogen at seven thousand metres were the same as at sea level. The principle was a simple one; air freed from water vapour and carbon dioxide was passed over heated copper which combined with oxygen whilst nitrogen passed into a weighed glass globe. The increase in the weight of copper yielded the weight of oxygen in the sample of air. If the principle of the method was elementary, the refinements were far from simple. Apart from the intricacies of the experimentation it was necessary to determine afresh the densities of oxygen and nitrogen as the known values at that time were found to be inaccurate. They concluded that air, free from carbon dioxide and water vapour, contained 20.81 per cent of oxygen and 79.19 per cent of nitrogen by volume. Boussingault did not lack experience in precise measurement and experimentation, but it is doubtful whether he had ever before conducted an experiment which demanded such meticulous care as that of measuring the composition of the air.

The possible variation in the amount of carbon dioxide in the air between town and country was the subject of a series of experiments in which simultaneous observations were made, with the help of a colleague, Lewy, in Paris and in nearby country districts, for example Andilly near the Montmorency forest north of Paris.[64] They found that town air contained a little more of the gas than country air but the variations in their results rendered their conclusion somewhat unconvincing. Boussingault queried whether the production of carbon dioxide in a town really made so much difference to the composition of the air.[65] An estimate of the amount of this gas produced in a day in Paris illustrates his statistical approach to the problem. Carbon dioxide, he assumed, was produced by people, horses and the burning of fuels. The population of Paris was then just over nine hundred thousand whose daily output of the gas could be calculated because his colleague, Dumas, had found that the average person respires three hundred and seventy litres per day. Boussingault had already found the volume of the gas respired by a horse,[66] but how did he know that there were about thirty-one thousand horses in Pairs? This figure was obtained from the weight of hay, straw and oats which passed through the gates of the city and on which city dues were paid. Finally, the toll statistics for the import of wood, coal, charcoal, oil, wax and tallow candles into Paris were incorporated into his calculations from which he concluded that in twenty-four hours in Paris there was a production of nearly three million cubic metres of carbon dioxide, in spite of which there was very little difference between the amount of carbon dioxide found in the air of the College de France and, what was then, the country air of Saint-Cloud. This result did not astonish him for if the twenty-four hour production of the gas was released then it would form a layer of about eight and a half centimetres' thickness and would be quickly dispelled by air currents.

Cooperation with Dumas at the *Sorbonne* meant that Boussingault was engaged in investigations on a number of substances derived from plant and animal tissues such as suberic acid from cork, and 'sugar of gelatine' or glycine from glue – the first example of an amino acid being prepared from a protein. There were also papers on bitumins but although these were related to the products of the petroleum fields of Alsace, and therefore to his income, he was no longer interested in these topics after 1840. From then to his appointment at the *Conservatoire des Arts et Mètiers* in 1846 his Sorbonne contributions were almost entirely limited to investigations on the composition of air. A perusal of Boussingault's scientific articles during the period 1835–1845 reveals that whereas there were many papers on South American subjects, crop husbandry and animal nutrition, there were few on pure chemistry and analysis which appear to have been necessitated by his appiontment at the Sorbonne. Those on suberic acid, glycine and the composition of the air lack the enthusiasm and the obsession of the Bechelbronn projects. He was serving two masters, himself and Dumas, and whereas the Sorbonne tasks were, perhaps, dictated by Dumas, his Bechelbronn investigations, as described in the following chapters, were certainly of his own volition.

CHAPTER 6

THE EXPERIMENTAL FARM AND CROP ROTATION

We must retrace our steps for a while and relive a portion of the South American scene. If during his nomadic life from 1822 to 1832 Boussingault ever had a *pied-à-terre* it was at Marmato in the valley of the rio Cauca whence came several of his articles on the analysis and treatment of ores. At the Marmato goldmine the particles of gold were embedded in a reef of iron pyrites, or iron sulphide, which was crushed and washed, the waste pyrites flowing into the river. Boussingault reasoned that on roasting the ore the heavy iron sulphide would be converted into the lighter iron oxide which would be washed away with greater ease.[1] But this meant that extra labour was required which could only be obtained from afar. In his *Mémoires* he recalled the problem in this way:

> The black population was insufficient for the work. We enticed labourers from the province of Antioquia; they arrived carrying with them rations for a fortnight, then returned home in order to come back again. To stabilise the labour force their food supply had to be assured. It was then that we put in order a large banana plantation on the farm Cucurusape on the banks of the Cauca. We encouraged land clearance for the sowing of maize, yuccas and vegetables. Trade with Antioquia soon brought wheat-flour, cocoa and coffee. It was through organising this tropical agriculture that I understood that we should demand from the soil the food necessary for the population. In other words, one must cultivate to live. My study of agriculture dates from that time.[2]

Although Boussingault ably describes this period as a turning point in his life, a revelation that his scientific path led in the direction of agriculture rather than in mining, there were other motivating factors. There was his observation of the important rôle of nitrogen exemplified by the lush growth of crops in the sterile sandy soils of Peru through the application of the locally mined *caliche,*[3] an impure sodium nitrate, or by the use of sea-bird manure or guano with its high proportion of ammonium salts.[4]

Marriage to a farmer's daughter now made it possible to pursue a career, if not yet as an agronomist but certainly as an agriculturist. At the farm Bechelbronn there was an association of rural activity and mining industry as in summer the miners were employed on the farm and they returned to the mine in winter where use was made of the farm wagons for transporting bituminous sand, coal and wood. If, during the winter, work ceased owing to poor mine ventilation, the mine labourers returned temporarily to the farm. Other sources of farm labour were to be found among retired miners, plot-holders without stock who liked working with animals, and, of course, the families of the proprietors. Agriculture was never sufficient to support the local population and some industry was necessary.[5] It appears remarkable that although Boussingault later inherited a feudal organisation and was in the midst of a socio-agronomic transformation he made no reference to it. When he first came to live at Bechelbronn, as a visitor in 1836, the residents consisted of the LeBel and Mabru households. Antoine Mabru was director of the Pechelbronn oilfield and was related by marriage to the LeBels. On the farm there was also a farrier, ploughwright, foreman, cowman, pigman, governess, four cooks and five domestics, altogether twenty-three people of whom half were Catholics, but toleration and independence were characteristics of Alsace, where families were amicably divided between Protestants and Catholics.[6] Others were present for meals, and because of the considerable fat content of the washing-up water which was supplied to the pigs, Boussingault undertook analyses of the water which provoked such extreme nausea that he complained to his colleague, Dumas, that he had lost appetite but not weight.[7]

The Influence of Albrecht Thaer

From 1836 to 1841 Boussingault conducted a series of classic crop rotation trials at Bechelbronn, assisted by his brother-in-law, Louis Frédéric Achille LeBel. Crop rotation has a long history. In the first century B. C. the Roman poet, Virgil wrote:

> . . . or, beneath another star, sow yellow corn in lands whence you have first carried off the pulse that rejoices in its quivering pods, or the fruits of the slender vetch, or the brittle stalks and rattling tangle of the bitter lupin.[8]

But despite its long history a distinction may be drawn between rotations prior to the eighteenth century, which included a fallow period, and later rotations where the fallow was replaced by an artificial pasture of lucerne which was regarded as a crop. This innovation has been credited to an English farmer, Jethro Tull who published a book on his ideas and methods in 1733 entitled *Horse-hoeing Husbandry*, in which he advocated constant ploughing as a substitute for manuring. He assumed that very small particles of soil could enter the pores of the roots, but however erroneous was this throry his remarks on crop rotation and artificial pastures render his book of historical importance. A French translation[9] had a profound effect on French agronomists, if not on French agriculture, for they became aware of the practicability of an almost continuous cultivation, made possible by a perpetual replenishment of the soil. Yet after the Revolution the practical results of this agricultural propaganda were extremely disappointing, for the new agricultural methods had been accepted by the leisured rich but not by the French peasant with his traditionally conservative outlook. Therefore, at the beginning of the nineteenth century crop rotation was part of the agricultural philosophy of the educated but was not practised by the mass of farmers; indeen, it has been said that the new agriculture in France "was less a fact of economic history than a fact of the history of ideas".[10]

There was another and nationalistic reason for the apathy towards English agricultural practice, a symptom of anglophobia arising from the critical attitude of the British towards the Revolution and the growing military power of France under Napoleon Bonaparte. Progressive methods were sought elsewhere and great interest was taken in the work of the German agriculturist Albrecht Daniel Thaer (1753–1828) at his agricultural institute at Möglin near Berlin. Boussingault's interest in Thaer and his works may have been an expression of their similar introvert natures. Thaer, son of a physician at Celle in Hanover, received a medical education at Göttingen and returned to his native town to practise, much to the consternation of the local medical fraternity who at first wondered whether there was sufficient room in the churchyard to bury his patients! On marrying, he acquired a 16 acre small-holding on which he grew crops, fruit and flowers, later introducing cattle. As his fame as a farmer spread throughout Europe he gradually withdrew from medicine. However, the occupation of Hanover by Napoleon's troops in 1803 meant that the Thaer family became refugees in Prussia until the king of Prussia appealed to him to start an agricultural school at Möglin, just outside Berlin. His later days were plagued by ill-health and his death in 1828 brought to an end an important chapter in the history of agriculture, noted for a pioneer attempt to quantify a way of life which hitherto had been mainly subjective in outlook.

Because Boussingault's work on crop rotation was an extension of Thaer's experiments, a summary of the German agriculturist's ideas is given. Thaer's interest in the related topics of crop rotation, soil exhaustion and soil fertility led him to formulate a scale of soil fertility. Instead of assuming the existence of a perfectly fertile soil which was gradually exhausted by successive crops he took the more practical view of supposing that a soil when so exhausted reached a state when it had what he called 'natural fertility', which was the fertility needing a fallow period to follow, or before an application of manure was necessary. This was a state below which no land should be allowed to deteriorate and he gave a value of 40 to this state of natural fertility, a kind of lower fixed point for soil. This implied that plants would still grow in it but there would be no cash return for the seed sown. Thaer did not attempt to describe the zero on his fertility scale but it would approximate to the incinerated sterile soils used later by Boussingault in his experiments on nitrogen fixation. Further values were given by Thaer; a land left fallow for a season would increase its degree of fertility by 10, and if followed by manuring there would be a further increase of 10 for every ton of manure added per acre. Thus a field with 'natural fertility' only and left fallow for a season and then given manure to the extent of five tons per acre would end up with a degree of fertility of 40 + 10 + 50 or 100. In general the growing of crops on this acre would reduce the degree of fertility and Thaer made estimates of the 'degree of impoverishment' due to the production of one bushel of each crop. With such figures Thaer endeavoured to demonstrate the losses and gains in soil fertility over a period of years.

The Crop Rotation Trials at Bechelbronn

As much as he admired the work of Thaer, Boussingault was not prepared to agree wholly with Thaer's statement that soil exhaustion was proportional to the quantity of nutritive substance found in the crops, for it was an assumption that all organic material in a plant was derived from the soil. By the end of the eighteenth century it was already known through the researches of de Saussure, Ingen-Housz and Senebier that a plant increased its organic content by the fixation of carbon dioxide in the presence of sunlight and the green material of plants, isolated in 1818 as chlorophyll. Their results were either accepted reluctantly or ignored for about forty years. It was, undoubtedly, an act of faith on the part of chemists to believe that the insignificant amount of carbon dioxide in the air could account for plant growth rathen than the thick pile of humus spread over the earth's surface, or at least over many parts of it. Moreover, there was the air, that "great sea of nitrogen" as Boussingault put it; was it merely fortuitous that an essential element of plant life bathed the very plants themselves? These were questions to which he was to give a large part of his professional life. He also disagreed with Thaer's belief that an alternation of crops would completely exhaust a soil. So impressed was Thaer by the nutrient balance in soil and crops that he advocated a succession of abundant and scanty crops. Not that he was unaware of the improvement effected in a soil by the growth of legumes but he attributed this to various reasons, the air which they exhaled, the stems and large roots which they left in the soil, the loosening effect on the soil due to their culture, or even the shade cast by their large leaves. Yet he did not remark on the root nodules which he must have seen, and which supplied the answer to later workers. However, in crop production, as in animal husbandry, Thaer found it necessary to express quantitatively the various processes involved and gave a quantitative basis to crop husbandry which was extended by Boussingault in terms of carbon, hydrogen, oxygen, nitrogen and mineral salts, basic materials of plant life.[11]

Because the crop rotation trials were started at Bechelbronn in 1836 the farm is recognised as the first agricultural experimental station and its centenary was celebrated on October 30th, 1936 in Paris by a lunch and a tour of the various buildings where Boussingault lived and worked, organised by the *Association française pour l'étude du Sol.*[12] As he was lecturing at the Sorbonne for half of the year he was assisted by his brother-in-law who probably supervised the important preparation of the soil before planting, e.g. the ploughing and harrowing, for Boussingault makes no mention of these operations. Rarely does he mention the chemist Letellier who assisted him in analysing the vast number of plant material specimens. Of the six rotations examined four of them were conducted at Bechelbronn and two on other farms. A typical rotation[13] and one favoured at Bechelbronn was as follows:

1st Year:	*Potatoes.*
2nd Year:	*Wheat*, sown in the autumn of the first year.
	Clover, sown in the spring.
3rd Year:	*Clover*, two further crops, the third crop ploughed in.
4th Year:	*Wheat*, in the clover land.
	Turnips, after the wheat.
5th Year:	*Oats.*

The manure applied to the fields consisted of the excrement of horses, oxen and swine; straw litter impregnated with urine; peat ashes, and gypsum (calcium sulphate). Manure was applied just before weedable crops were

Fig. 12. Meeting at Bechelbronn of l'Association Internationale de la Science du Sol, July 7th, 1934 (Archives de l'Académie des Sciences, Paris).

planted such as potatoes and turnips as it was too difficult to weed a field of broadcast wheat, oats or clover.

Except for their importance, the results of five years of soil preparation, manuring, planting, cropping and chemical analysis seem all too brief. It was found that the weights of carbon, hydrogen, oxygen and nitrogen in the crops removed from the fields were greater than those elements supplied by the manure. Only the mineral salts had decreased but these were obviously in the soil. Yet carbon fixation by the leaves and the nitrogen increase due to the legume crop were not sufficient to enable the rotation to be repeated without the assistance of another application of manure. At first no account was taken of the crop residues left in the soil such as roots, stubble and the haulms of potatoes but a separate investigation showed that crop residues returned to the soil a weight of essential elements equal to one-half of those found in the original dressings of manure.[14]

The main results of an examination of the five years of rotation trials can be briefly summarised:

> A balance sheet of farming operations in terms of essential elements had been drafted.
>
> A legume crop causes an increase in the nitrogen content of the soil.
>
> If a land is constantly well-manured there is no necessity for a rotation to be followed, indeed a successful monoculture may be practised.

Boussingault's third, and over-optimistic, statement was the result of a study of Jerusalem artichoke monoculture, so popular in Alsace, where the gain in elements is mainly confined to the large stem which is economically useless. There are exceptions, however, where a well-manured rotation does not succeed especially if successive rotations contain clover. The second statement can be regarded as the first major discovery of Boussingault's career yet the biochemical reasons were to elude him to the end of his life.

Boussingault's memoir on crop rotation was examined by a committee of the *Académie des Sciences* including Thenard and Dumas who said that Boussingault, by the use of the balance, had dispelled doubts about the results achieved, which could not be said of previous investigators.[15] Yet such high praise does not conceal the committee's lack of appreciation that a long term farming operation, beset by possible seed germination failures, inclement weather, pests and diseases, had been elevated to the status of a precision

ASSOLEMENT N° 1.

ANNÉES.	SUBSTANCES	RÉCOLTES par hectare.	RÉCOLTES sèches.	CARBONE.	HYDROGÈNE.	OXIGÈNE.	AZOTE	SELS et terres.
		kilogr.	kilogr.	kilogr.	kilogr.	kilogr.	kilogr.	kilogr.
1re	Pommes de terre................	12800	3085	1357,4	178,9	1379,0	46,3	123,4
2e	Froment..........................	1343	1148	529,3	66,6	498,2	26,4	27,5
	Paille de froment...............	3052	2258	1093,0	119,7	878,2	9,0	158,1
3e	Trèfle (en foin).................	5100	4029	1909,7	201,5	1523,0	84,6	310,2
4e	Froment..........................	1659	1418	653,8	82,2	615,4	32,6	34,0
	Paille de froment...............	3770	2730	1350,4	147,8	1085,3	11,2	195,3
	Navets dérobés..................	9550	716	307,2	39,3	302,9	12,2	54,4
5e	Avoine............................	1344	1064	539,5	68,0	390,5	23,3	42,6
	Paille d'avoine..................	1800	1283	642,8	69,3	500,4	5,1	65,4
	Somme............................	40418	17791	8383,1	973,3	7172,9	250,7	1010,9
	Engrais employé.................	49086	10161	3637,6	426,8	2621,5	203,2	3271,9
	Différence........................	....	+ 7630	+ 4745,5	+ 546,5	+ 5551,4	+ 47,5	− 2261,0

Fig. 13. Rotation no. 1 at Bechelbronn, from *Annales de chimie et de physique*, 1841, Vol. 1, pp. 208–246 (Masson).

experiment. They did admit that "the characteristic feature of M. Boussingault's researches is to have introduced the balance into the study of the problems of plant physiology". Briefly, he had taken the balance to the field.

Criticisms of the Rotation Trials

Such iconoclastic conclusions were bound to invite criticism. Justus von Liebig was one of the outstanding chemists of the nineteenth century and certainly the pre-eminent chemist of his day. To his laboratory at Giessen came students from France, Great Britain and the United States who later returned with the enthusiastic dedication of those who have studied under a master. His research and teaching methods were thus advantageously spread far afield. But whereas his researches and discoveries were founded on strictly well-controlled and oft repeated experiments, his hypotheses on agriculture and medicine were often the products of his lively intuition. Yet his great contribution to agriculture was the publication in 1840 of *Chemistry in Its Applications to Agriculture and Physiology* which, although it espoused implausible theories, provoked considerable thought among farmers concerning the scientific aspects of their calling. His polemic style of authorship probably made a greater impact on agriculture than any other agricultural work of that time.

Not unexpectedly, criticism of Boussingault's rotational trials and conclusions came from Liebig as a result of an experimental dichotomy which had arisen. Whereas Boussingault relied on the results of crop trials extending over five or more years, Liebig placed most emphasis on analyses of plant ashes, for these, he believed, gave him an indication of the nutritional needs of a plant. Thus the results of crop rotations were being viewed from two completely different standpoints. That the nitrogen content of the soil was important was a matter which could be almost completely ignored by Liebig with his obsession about ammonia in the atmosphere providing the nitrogen needs of a plant. Ash analysis, he thought, yielded the answer to soil fertility because it revealed the nature of the mineral salts necessary to acquire and maintain fertility.[16] The ash analysis approach was limited in its usefulness by the elementary state of quantitative analysis, and even Liebig admitted that "everyone who has experience of this kind of analysis is aware that a quantity under one per cent cannot be determined with accuracy".[17] Briefly, Liebig considered that the key to fertility was to be found in the mineral content of the soil and fertilisers, whilst Boussingault insisted that the nitrogen content was paramount in importance but recognised the value of mineral salts. Liebig admitted the necessity for maintaining a rotation of crops but arranged the crops according to the dominant mineral salt which each crop removed from the soil. Thus mangel-wurzels, turnips and corn were potash plants; legumes and tobacco were lime plants; cereals were silica plants, and it followed that soil fertility was restored as a result of implementing the findings of ash analysis. Liebig appeared unable to recognise the specialised nitrogen producing[18] function of legumes and attributed any nitrogen increase to the application of manure without realizing that a cereal following a legume removes nitrogen produced by the legume in the soil. Moreover he did not appreciate that only part of the nitrogen in manure is immediately assimilable by the plant, the rest undergoing decomposition to an assimilable form. The philosophical attitudes of the two men differed, for whilst Boussingault suspended judgement on the necessity for rotations, Liebig thought he could explain in purely chemical terms the transformations that took place in the soil as opposed to a vitalistic explanation, common at the time, which endowed living material with a capacity to effect chemical changes unknown to the inorganic world. Yet it must be said that Boussingault never, at any time, believed that chemical changes in plants or soil could be explained in extra-scientific terms. Debate in polemical terms with Leibig was shunned by Boussingault except to comment on the belief that the real value of manure was in the mineral salts and not in the nitrogen, which provoked the only known example of Boussingault sarcasm when he interpreted Liebig's beliefs as an invitation to all farmers to burn their manure heaps![19]

The debate on crop rotation, initiated by Liebig, met a warm response from John Bennet Lawes and Joseph Henry Gilbert as Rothamsted in England. Lawes had started pot experiments on fertilisers in 1835 but by 1843 the increase in these investigations necessitated some assistance and thus started a long period of cooperation with Gilbert, a former student of Liebig at Giessen. Lawes and Gilbert had doubts about Liebig's assertion that mineral salts were of paramount importance as Lawes had recorded great success with nitrogenous phosphate fertilisers from 1840 onwards.[20] Consequently Lawes enquired whether a plant needed only the substances found in the ash; he regarded such an investigation as perhaps the most important contribution which chemistry could offer to agriculture.[21] The answer was given by the now famous Broadbalk field at Rothamsted which was divided into test plots for wheat, turnips and legumes, with various fertilisers including an invention of Liebig, 'Liebig's Patent Manure', which contained the mineral salts essential for growth. The soil had been exhausted by a four-year rotation without the application of manure or fertiliser, similar to Thaer's state of 'natural fertility'. Lawes and Gilbert, by a long series of carefully planned and meticulously supervised trials, demonstrated that mineral fertilisers alone were insufficient for the growing plant, and that the loss of nitrogen in the soil was the important factor in its exhaustion. It was not as Liebig had declared in a much publicised statement:

> The crops of a field diminish or increase in exact proportion to the diminution or increase of the mineral substances conveyed to it in manure.[22]

Further, the result of wheat trials conducted by Lawes and Gilbert, announced in 1847, showed that the application of farm manure produced a better crop than the ash of the same quantity of manure or by the use of mineral salts,[23] thus contradicting Liebig's views. This result had already been suggested in a series of analyses of fertilisers by Boussingault and a colleague, Anselme Payen, in 1841–1842 where the value of the manure or fertiliser was found to depend on the amount of nitrogenous organic matter that it contained.[24] That all plants required at least nitrogen explained why a fertiliser could be applied to a variety of crops which might have needed specific fertilisers according to the nature of the crop, the soil and the climate.

Encouraging the use of fertilisers among the financially shrewed farmers of Alsace was no easy matter, as Boussingault found. The farmers obtained a good price for their wheat and were not inclined to increase their yields by a cash outlay on fertilisers. A chemical fertiliser was available in France and manufactured by heating rock phosphate, or calcium phosphate, with tar. This was the Buran process in which the quicklime, produced from limestone associated with the rock phosphate, caused the particles of rock phosphate to break up when moistended in the soil and become more effective. In Britain, rock phosphate was treated with sulphuric acid to form 'superphosphate' which was a patent of Lawes and produced in his fertiliser factory at Deptford, London from 1843. In a chauvinistic episode at the *Société Centrale d'Agriculture* in 1864 Boussingault enquired why the English continued to use this expensive method and was told that they were like sheep and did not trouble to find a more economic method.[25]

The rotation experiments coincided with the start of family life. First a daughter, Berthe Gabrielle, was born in 1836 at the Château LeBel and in the register of births at Lampertsloch Boussingault described himself as a Chevalier de Légion d'honneur, and, surprisingly, dean of the faculty of Sciences at Lyon, a post which he had left fourteen months previously! The

Marie Joseph Achille
LeBel (1772–1842)

Louis Frederic Achille
LeBel (1807–1867)

Joseph Achille LeBel
(1847–1930)

Fig. 14. The LeBels – from *Historique de Pechelbronn: 1498–1918* by Paul de Chambrier (éditions Victor Attinger, Neuchâtel, 1919).

second daughter, Alice, was born in Paris in 1839 and the only son, Joseph was also born in the Château LeBel in 1842 when Boussingault described himself as a member of the Institute of France.[26]

Boussingault Publishes an Agricultural Text

He was now conscious that his experiences in South America and at Bechelbronn merited a book embodying his experiments, observations and theories, as well as those of others in the same field. The first intimation of such a work is found in a letter to Dumas[27] who had expressed a suspicion that his friend was writing an agricultural book. Boussingault's long reply described his reasons for writing the work, *Economie rurale*, the general plan and his difficulties, but also revealed his anxieties regarding its possible success and wondered whether to publish lest the general public should regard his ideas as those of a lunatic. His original intention to combine a series of already published memoirs in book form[28] was abandoned which was fortunate as it would then have been suitable for scientists but not farmers. The work eventually consisted of two volumes each of over six hundred pages. In the above-mentioned letter to Dumas he mentioned the difficulties of writing an authoritative text because he could not ignore the smallest fact which he considered important and must therefore consult research data and scientific papers which, hitherto, had not been familiar to him, a surprising statement for surely these difficulties must have been present when writing his two Sorbonne theses for a doctorate. He was worried lest Dumas might mention the book to others for he alleged that "strange questions are being asked on the banks of the Rhine", a reference to possible scientific espionage by Liebig's workers. What is more, when a German translation was contemplated he implored Dumas to keep secret an interview concerning a possible German translation for he was convinced that Giessen spies had discovered his current analysis programme – first bean ash then hay ash.[29] Does one misjudge the seriousness of the situation where ash analyses are of international importance? Yet Liebig in a very cordial letter to Boussingault in May 1844[30] offered to have *Economie rurale* translated into German although it is not known whether this was the translation published in 1844–45 in Halle.

Boussingault's letter to Dumas, referred to above, contained an outline of the first five chapters, and eventually the ten chapters were, (i) The Physical Phenomena of Vegetation, (ii) The Chemical Constitution of Plants, (iii) Vinous Fermentation, (iv) Soils, (v) Manures, (vi) Mineral Manures, (vii)

Crop Rotation, (viii) Feeding of Animals, (ix) Stock in General, (x) Meteorological Considerations. He intended including a chapter on 'The General Principles of Agricultural Book-Keeping' but this was omitted – a pity because an economic survey of the farm Bechelbronn showed that it was being run at a loss, or rather it was being subsidised by the oil revenues of Pechelbronn.

It is difficult to envisage the readers for whom Boussingault was writing; the agricultural worker was illiterate and his employer had little knowledge, if any, of chemistry. An English translation, *Rural Economy*, appeared in 1845 but it is difficult to agree with the translator who considered the book to be "intelligible and instructive to every agricultural enquirer, however superficial his acquaintance may be with the details of chemical science", yet he was compelled to delete portions where "the author appeared to be forgetting the rural economist for the mere chemist". A very limited group of agriculturists could have had the scientific background enabling them to benefit from, or even understand this work, and perhaps the answer is that it was written for those future generations of farmer students who were to attend his winter lectures at the Conservatoire des Arts et Métiers from 1845.

It would be laborious to describe the two volumes item by item for if one term suitably classifies them it is 'encyclopedic'. Some information is irrelevant such as the preparation of suberic acid from birch bark, or an explanation of the temperature when water is at its maximum density. The first volume includes some interesting and simple soil tests for tenacity, imbibition by water, tendency of soils to become dry, and soil contraction. It closes with an exciting account of how an engineer, Nicolas Brémontier, started in 1787 to halt the advance of sand dunes around the Gulf of Gascony by planting pine, broom, oaks and willow. Boussingault regarded this as an example of the farmer's struggle against the elements.

Early in the second volume there appear the terms *terreau* and *terre végétale* whose literal translation into English would not only prove incorrect and misleading but would hinder an understanding of Boussingault's work on soil and soil nitrification. A certain ambiguity arises when dictionary definitions are sought for *terreau* appears to mean 'mould' or 'humus', and the English translator warned the reader against regarding all three terms as synonyms. Confusion ceases when it is understood that Boussingault, in using these terms, was favouring an analytical and dynamic interpretation of soil rather than the familiar stratigraphical and static distinction between sub-soil, top-soil and humus. Fortunately he gave a definition of *terreau*:

Dead material, once living, such as wood, straw, leaves, exposed for a long time to the action of air and water, and by being transformed into a brown, almost black, substance when damp, becomes powdery when dry.[31]

In correlating English and French terms any definition could be debatable, but it is suggested that a chronological explanation might apply, for example, dead plant material or *résidus des plantes* becomes *terreau* in course of time, which in a more advanced stage of decomposition becomes *humus*. Humus mixed with soil becomes mould or *terre végétale* which is characteristic of the upper layers of top-soil when under cultivation. And, of course, when such a *terre végétale* receives manure and nitrification takes place with the production of saltpetre there are present the conditions necessary for a *nitrière* or saltpetre-bed.

Boussingault in 1843 subscribed to the prevalent view that manure should be ploughed into the soil when fresh, the ammoniacal odour of dung giving rise to the belief that manure on standing lost some of its fertilising power. This was shown to be incorrect by a German chemist, J. C. A. Voelcker, settled in England, who found that not only was the amount fo free ammonia in manure relatively small but that the nitrogen content on standing depreciated very little.[32] Only two pages of *Economie rurale* are devoted to the subject of compost for which he then had little enthusiasm and merely conceded that mixing manure with plant debris distributed the manure more evnely. Yet Boussingault refused to believe that more fertiliser was created in this manner for he could not imagine that plant debris, which he termed 'inert', could be activated by the presence of manure. There is more than a hint of vitalism here, the dead plant debris being thought incapable of assuming the lively character of manure. Later, he became a fervent convert to composting and was found, in 1864, defending the practice at the *Société Centrale d'Agriculture* against the opinions of chemists.[33]

Perhaps the most valuable treatment of *Economie rurale* is to compare it with Thaer's *Principles of Agriculture* to which it bears a superficial resemblance. An examination of the chapter and section titles of both works reveals that their authors endeavoured to cover the whole field of animal and crop husbandry. Thaer omitted references to the chemical phenomena of vegetation and the inorganic materials contained in plants: he neglected the subject of the chemical constitution of vegetable substances and did not treat of fermented and spiritous liquors, nor of meteorological considerations, all of which were included in *Economie rurale*. On the other hand, Boussingault's book was wanting in its exclusion of such subjects as the purchasing or leasing of a farm; capital and labour; keeping farm journals and registers; and questions of practical tillage which were to be found in *Principles of Agriculture*. Whereas Boussingault wrote a treatise on the science of agriculture it can be said that Thaer dealt with the practice of agriculture as a means of producing food and earning profits. Boussingault's work was an indication of the increasing understanding of the phenomena of crop and animal husbandry which had arisen in the second quarter of the nineteenth century, to which he had made a significant contribution. The differing approaches arose because Boussingault was a scientist whereas Thaer was a farmer.

If there was little progress in the sphere of crop rotation from 1809, when Thaer published his results, to 1838 when Boussingault first reported his preliminary findings, it was not due to lack of interest but it was a period of consolidation during which theories of crop rotation were suggested and examined. They were either physical or chemical and the one favoured by Boussingault was that of differential root length where nutrients at different levels of the soil were removed by successive crops in a rotation.[34] Another theory suggested that crops which attracted insects must be followed by crops which have no such attraction. If insects are a nuisance then so are weeds. If cereals when broadcast render weeding difficult, then they should be followed by a weedable crop[35] such as potatoes or mangel-wurzels. Chemical theories were of two kinds, the first postulating the absorption of nutrients by roots in a selective manner.[36] The other was that of root excretion where the plant root absorbed soil nutrients unselectively and rejected those for which it had no further use.[37] A variation of this theory was that the plant roots excreted materials which were toxic to that type of plant but to no other, hence a rotation was imperative.[38] Although Boussingault gave great prominence in his publications to crop rotation theories, his experiments at Bechelbronn were not designed to test any of them.

What had Boussingault achieved? Was it merely a debit and credit balance sheet for the arable portion of Bechelbronn over a period of five years or more? Was it only the discovery that the planting of legumes increases the nitrogen content of soils or that plant debris contributes much to the soil's fertility? Independently they are important agricultural achievements but together they exemplify that the whole is greater than the sum of the parts. Boussingault was implicitly piecing together the parts of what was later referred to as the nitrogen cycle in nature, the nitrogen compounds in plant and animal waste enriching the soil which yields such compounds to the plants whose debris return them to the soil, or by way of the animals which eat the plants. The inter-relationship between the nitrogen of animals, plants and the soil was clear in its outlines. But what of "the great sea of nitrogen" – the atmosphere? There were to be many years and many disappointments before the vegetative function of atmospheric nitrogen was elucidated – but meanwhile, Boussingault was also busy with a classic problem in animal nutrition.

CHAPTER 7

THE EXPERIMENTAL FARM AND ANIMAL NUTRITION

For many centuries farmers had amassed an empirical corpus of facts about stock-feeds and there was much practical knowledge about the value of such feeds as hay, lucerne and turnips. Interest in animal nutrition increased with the greater use of stall-feeding in winter, the availability of a greater variety of stock-feeds, and with the invention of the Bramah press in 1795 with the consequent use of oil cake with a high nitrogen content. The development of the subject of animal nutrition has progressed though the predicatable stages of description, generalisation and hypothesis, although not always in that sequence, there being repeated interaction between these stages. Perhaps the first advance in the generalisation stage of animal nutrition was the discovery in 1816 by François Magendie that dogs could not survive on a diet emtirely lacking in nitrogenous substances.[1] Parallel to this development in the value of stock feeds was a recognition of nutrient groups so that foods could be classified according to their content, an advance mainly due to William Prout who, in 1827, said that in examining foods it was necessary to assess the "saccharine, oily and albuminous divisions" of the food,[2] referring to what is now known as the carbohydrates, fats and proteins.

Among the first comparative studies of stock-feeds were those of Albrecht Thaer who published the results of his investigations between 1810 and 1812. Noting that cattle, horses and sheep could be kept in excellent condition by feeding tham with good hay, he partially substituted other foods such as potatoes, turnips, clover-hay and swedes in order to discover the weights of these foods which were equivalent to his standard of 100 parts of hay and thus obtained a table of 'hay equivalents' in which 100 kilos of hay was equivalent to 200 kilos of potatoes, 266 kilos of carrots and 90 kilos of dry lucerne. Thaer's equivalents became highly controversial because hay is a stock-feed of variable nutritive value according to age. Also, he placed great emphasis on the quality of the dung produced by various feeds and insufficient stress on the assimilation of nutriment. This is understandable in an age before the use of chemical fertilisers when the dung producing capacity of the farmer's stock was of prime importance as the sole source of fertiliser.[3]

Boussingault's Wheat-Flour Equivalents

Boussingault, in the eighteen-thirties was also seeking a standard by which stock-feeds could be compared. He was aware not only of Thaer's hay equivalents but also of Magendie's investigations on the importance of nitrogenous foods, and concluded that the nutritive value of a foodstuff was proportional to its nitrogen content which he then determined for the various stock-feeds examined by Thaer. His method, due to Dumas, consisted of heating a sample with copper filings and measuring the volume of nitrogen given off. There was found a distinct correlation between Thaer's table of hay equivalents and Boussingault's table of nitrogen content. Boussingault then decided to make wheat flour his standard material which proved to be less variable in quality and therefore a more stable standard of comparison. Having determined the nitrogen content of wheat flour he was able to relate stock-feeds as equivalent to 100 kilos of wheat flour, e.g., if the wheat flour equivalent of potatoes is 613 then 1 kilo of wheat flour has the same nitrogen content as 6.13 kilos of potatoes.[4]

If the nutritive value of a stock-feed corresponded to the nitrogen content, what was there in the stock-feed which contained nitrogen? Fourcroy in 1789 had shown that there was nitrogen in the gluten of wheat, the sticky substance left when starch is washed out of wheat flour.[5] Boussingault, having grown 24 varieties of wheat from all over the world in the *Jardin des Plantes* at the *Muséum d'Histoire Naturelle*, wished to find a rapid method of determining the gluten content of these wheats. He was seeking a method by which the nitrogen content of the flour, obtained by Dumas' method, could be divided by a factor which would give the weight of the gluten. An analysis of gluten showed that it conained 16% of nitrogen, therefore if a flour was found to contain 4.8% of nitrogen then the gluten content was 4.8/0.16 or 30% His own Bechelbronn wheat which normally yielded a gluten content of 17.3% increased this value to 26.7% under the influence of the well-manured soil of the *Jardin des Plantes.*[6]

Having established that the value of a stock-feed depended on its nitrogen content, his animal nutrition trials of 1838 proceeded along two different but related paths which concerned first the quantity and composition of the milk yielded by cows receiving varying but equivalent feeds. The composition of the milk was expressed in terms of casein, butter, milk sugar, mineral salts and water. Secondly, did the animal, in its nutrition, make use of atmospheric nitrogen? To answer the first question as to whether variable but equivalent feeds influenced the milk vielded by the cow he fed eight cows with different but equivalent mixtures selected from hay, mangel-wurzel, potatoes, green clover, turnips, Jerusalem artichokes and straw. He concluded that the nature of the feed did not affect the volume or the composition of the milk, provided that the feeds were of equivalent nutritive value, i.e., their nitrogen content was the same.[7]

Do Animals Made Dietetic Use of Atmospheric Nitrogen?

It is not surprising to find that Boussingault questioned whether animals consumed the nitrogen of the atmosphere, for he reasoned that the quantity of living matter on the earth depended on the amount of nitrogen available, and beyond plant and animal debris the possible source of dietetic nitrogen was the atmosphere. The probable use of nitrogen by animals and plants was a recurring theme in his scientific papers. There was then the growing awareness that animal food had a dual function, that of maintaining the body at a constant weight and degree of health, or the maintenance ration, and also for the production of milk or for fattening, known as the supplementary ration. The nitrogen content of the food consumed by a cow over a period of three days was compared with that of the milk and excretions with the result that he concluded that there was no evidence of a dietetic nitrogen intake.[8] A similar experiment with a horse confirmed this result.[9] Had the cow and horse made direct use of atmospheric nitrogen, the nitrogen cycle in nature would have been considerably simplified, but another fifty years were to pass before the devious route was discovered by which atmospheric nitrogen is eventually digested by animals. Boussingault's emphasis on the paramount importance of stock-feed nitrogen was soon to be challenged, this time by Lawes and Gilbert, at Rothamsted as a result of their trials from 1847 to 1852, who found that the demands of respiration merited a supply of non-nitrogenous food.[10] Although the importance of nitrogen in a stock-feed was never disregarded, Boussingault's work on animal nutrition was greatly amended by others; henceforth his main contribution in animal husbandry was concerned with fat formation and he was never again to return to stock-feed analysis.

How Do Herbivores Produce Fat?

There commenced in 1843 a fierce controversy about the formation of fat in herbivores which include such farm animals as the cow, the horse, sheep, ducks and geese; Carnivores prey on herbivores but how do herbivores obtain fat from plants? German chemists, with Liebig as an effective propagandist, believed that a herbivore produced fat from other food substances, especially from carbohydrates. Their conclusions, later proved correct, were based on the most superficial evidence. French chemists, among whom were Dumas and Boussingault, believed that the fat of the herbivore already existed in the plant, a theory first proposed by two German chemists, Friedrich Tiedemann and Leopold Gmelin in 1825. The controversy owed its origin to a lecture by Dumas in 1841 when closing his course of lectures at the Ecole de Médecine which was afterwards published under the authorship of Dumas and Boussingault entitled *The Chemical and Physiological Balance of Organic Nature*, this being the English translation of the original title, *Essai de statique chimique des êtres organisés.*[11] In it they stated that their experiments had shown that fats are formed by the plant and not by the herbivore; further, they stated that fats passed unchanged into the animal to be burnt in developing heat, or fixed, more or less modified, in the tissues to serve the purpose of respiration. They concluded that the only material that would replace adipose tissue was ready-made fat supplied by plants. They were certain that the butter of cow's milk already existed in the tissues of the plant. Boussingault repeated these statements in his *Economie rurale*, published in 1843.

Liebig replied in 1842 saying that it was most likely that fat was formed in the herbivore from the carbohydrate in its feed.[12] Liebig also believed that animals formed fat from materials in the blood, but what was not known was how fats in the digestive system were changed, if at all, to make their way into the venous system. A knowledge of the mechanics of the transfer of fat globules was necessary, and although fats could be changed, or saponified, in the laboratory, the reagents used were much stronger than the aqueous solutions found in the digestive system. It was Claude Bernard who in 1849 discovered that the pancreatic juice decomposes fats into its constituents, a fatty acid and a glycerol, both of which could pass through the walls of the small intestine;[13] yet even this view was not accepted until the discovery of the pancreatic enzyme lipase.

A hypothesis, firmly held by the scientific community of the time, may be challenged by the emergence of an experimental result which cannot be fitted into the existing scheme of accepted belief. A crisis state is thus developed which in time is resolved by further experiment and an adjustment to the hypothesis. The diametrically opposed views of the French and German chemists on the subject of fat formation in herbivores caused such a crisis state in the young science of animal nutrition, not without its undertones of nationalism, which gave rise to remarks from both sides more worthy of the political than the scientific arena. Liebig had noted the amount of food given to two of Boussingault's cows, and also the nature of the excretions, after which he re-interpreted them in terms of another set of results obtained from cows at Giessen. Thus Dumas was caustic in his criticisms and accused Liebig of having drawn conclusions from "a truly imaginary cow which should have eaten at Bechelbronn the hay from Giessen; who, eating as the first cow, had supplied the excretions of the second cow and the milk of the first".[14] Such was the depth to which the debate deteriorated that Anselme Payen, a colleague of Boussingault, suggested that oil from maize had become rancid because it was grown in Lidebig's garden![15] Overriding such unworthy and petty remarks was the inaccurate analytical technique of the time, when Magendie could quote a case where some horses gave more fatty material in their excrements than that supplied in their food. Yet Magendie rightly insisted that it was one thing to recognise similarities between plant and animal substances but quite another to insist that animal tissues were formed exclusively from plant substances.[16] However erroneous were some of the theories of the French and German chemists, they introduced a dynamism into the subject by a study of chemical changes taking place within the body.

The theory of fat formation from carbohydrates, favoured by the Germans, had evolved from experiments conducted in 1780 by a Swiss entomologist, François Huber who fed bees in an enclosed hive solely on sugar and honey, and as a result a fatty substance, wax, was produced.[17] These results were later confirmed in 1842 by a German entomologist, Ferdinand Gundelach,[18] but metabolic differences between bees and cows did not render convincing the theory of fat formation from carbohydrates. So Liebig arranged for a cow belonging to Herr Koch, a butcher, of Giessen to be fed on oats and potatoes for six days. The excretions yielded the same weight of fatty material as was contained in the food, yet the cow gave as much milk as those of Boussingault at Bechelbronn, and so Liebig was convinced that the fatty subatances in oats and potatoes contribute nothing to the production of butter.[19] Then why were these opposing views held on either side of the Rhine? Their contradictions were the result of different feeding methods; whereas the Germans supplied unlimited rations, the French confined their trials to limited rations from which very little fat formation was possible. Even Boussingault noticed that outside the trials, on normal feeds, his pigs registered a more rapid increase in weight.[20]

It may have been symptomatic of the variety which he sought in his scientific life that he began to tire of the subject of fat formation. He wrote to Dumas that it was time he finished, that he was there to kill, to massacre, to skin and to boil, with blood flowing in cascades. He compares Bechelbronn with Spain,[21] referring no doubt to the bull-ring. He was also coping with foot-and-mouth disease but at last he decided to publish his results on the fattening of pigs, which at the same time represented a reversal of belief. No longer did he agree with the position held by his colleagues, and had given up his slogan, "Nothing fattens like fat".[22] In his anxiety to complete these experiments Boussingault spent part of the 1844–1845 winter at Bechelbronn which was most unusual as his winters were normally spent in Paris.[23]

Jean Persoz and His Geese

To clinch the matter it was necessary to use an animal which fattened rapidly and this was realised by a professor of chemistry at Strasbourg, Jean Persoz, who took note of his evnironment, that of the pâté de foie industry. Geese are penned and girls known as *gaveuses*, or 'stuffers', are employed to fill their beaks with maize. The geese are given an unlimited ration, unlike the limited feeds of Boussingault's pigs. With such an example before him it is not surprising that Persoz chose the goose as the subject of his experiments. He also profited from the local belief that a goose should be slaughtered from 18 to 24 days after stuffing has commenced.[24] That there was cooperation between Boussingault and Persoz is revealed in a letter from Boussingault to Dumas[25] in which he mentioned that he had prepared cages for Persoz' geese but had not the courage to undertake the work, and was much in need of a rest. "I shall become stupid if I persist in being stubborn in these queries. I have decided to let them take their course".[26] It was during these investigations that Persoz discovered the widely varying oil content of maize according to its age, and therefore much acrimonious discussion between French and German chemists could have been avoided if this had been known in 1843. This was not the only difficulty in assessing the fat content of a stockfeed such as maize, for what was urgently required was a practical definition of 'fat'. To the French chemists it was that part of a material soluble in ether, but Liebig objected to this criterion and showed that ether not only dissolved fat but also chlorophyll and a wax present in hay. All the chlorophyll and wax were found in the excreta and therefore did not play a rôle in fat formation.[27] Liebig was endeavouring to distinguish what today would be referred to as 'natural triglycerides', such as the tristearin of lard, from the whole group of fatty and related substances classed as 'lipids' which includes the waxes.

Persoz found that fat was produced in his geese at the expense of starch and sugar in maize; he also found that the blood of geese became richer in fat during the fattening process. Ten geese were used in his investigations and each had eventually to be dissected, the fat around the intestines removed and also the sub-cutaneous fat. So Justus von Liebig's intuition was confirmed, though the term 'intuition' implies no disparagement but is merely a reminder that his theory of fat formation in a herbivore was based on superficial and inadequate evidence. It was Boussingault who conducted the marathon series of experiments on fattening in which he made use of cows, pigs, ducks and geese. To Persoz, however, must be given the final credit, and Boussingault admitted that Persoz had priority in this discovery that fat in herbivores was formed from carbohydrates.[28] Boussingault's pigs had pointed the way but the goal had been reached by Persoz' geese.

In the subject of animal husbandry what had been achieved? Firstly a beginning had been made in evaluating stockfeeds in terms of a nutritive element, in terms of the nitrogen content, involving a liaison between farm and laboratory, between agriculture and chemistry. Secondly, an experimental beginning had been made on the problem of fattening and milk production, not just a mere increase in weight but, in the words of Lawes and Gilbert referring to their own experiments, "a source of saleable animal increase".[29] These results had been achieved with animals which were genetically inferior to those of today, or retarded through deficient feeding. In one set of records Boussingault mentioned that his cows averaged six litres of milk per day which would give a total[30] of 1860 litres per year allowing for a lactation period of 310 days. This figure compares unfavourably with the Friesland average of over 4000 litres. Pigs at Bechelbronn were not ready for fattening until they were 52 weeks old and this was sometimes extended to 70 weeks,[31] whereas today slaughtering ages are from 16 weeks for porkers to 28 weeks for the heavy hogs used for hams. Some of his pigs were slaughtered at 80 weeks and weighed 136 kilos whereas today such a live weight would be attained in 36 weeks, yet Boussingault's pigs were superior to those recorded by Liebig whose pigs weighed 35 kilos after a year with a fully fattened pig weighing no more than 75 kilos.[32]

Bechelbronn the First Agricultural Experimental Station

With the termination of large-scale agricultural experimentation in 1848 Bechelbronn ceased to be an agricultural experimental station and reverted to its former status as a farm. It was Adrian Gasparin whose varied career culminated in the directorship of the *Institut national agronomique* at Versailles, who remarked that Bechelbronn was the first of such stations.[33]
Yet this statement requires some critical examination because one has to distinguish carefully between an experimental station, an experimental farm, an agricultural college or institute, a farm school and a model farm. An agricultural experimental station is essentially a research station where investigations are undertaken in arable or pastoral farming accompanied by suitable examination of various aspects of the environment, for example, soil, manures and fertilisers, seed, crops, crop wastes, plant and animal diseases and the production of meat, milk, eggs, fruit and vegetables. It has the necessary facilities for the experimental examination of such items. It is not a school except for post-graduate study. Bechelbronn fulfilled these conditions but if these were doubts about its priority they would be raised by Thaer's establishment at Möglin where his friend Heinrich Einhof, as agricultural chemist, examined the farm materials in his laboratory and so a claim might be made that Möglin evolved the nucleus of an agricultural experimental station. Yet in the case of Bechelbronn there is no doubt that it started in 1836 as an experimental station in the modern sense of the term having priority in time over Rothamsted in England, opened in 1843, and Moeckern in Germany in 1852. Boussingault was certainly regarded by his contemporaries as the pioneer agricultural chemist as was confirmed at the Congress of Agricultural Chemistry held at Leipzig in 1877 when a telegram was sent to him, "Boussingault of Bechelbronn, to the first agricultural chemist his successors give a hearty cheer".[34]

An interesting comparison can be made between the fortunes of Bechelbronn and Rothamsted; both were originally private estates but as they developed there was no attempt to form a Bechelbronn Trust as happened at Rothamsted. Boussingault appears to have had more than the average share of personality difficulties when working with others, but strong personality differences did not prevent excellent professional cooperation between Lawes and Gilbert at Rothamsted who had great respect for one another. The reactions of the local farmers illustrate, perhaps, different national attitudes; the Alsace farmers in 1848 sent Boussingault as their deputy to the *Assemblée Nationale* but the Hertfordshire farmers subscribed to build the Rothamsted Institute in 1855. Lawes and Gilbert were better organisers than Boussingault but their superior management was necessitated by the different approach to animal husbandry between Rothamsted and Bechelbronn, indeed between Britain and the Continent. Whereas at Bechelbronn emphasis was on the character of the forage and the comparative value of different foods, in Britain prior concern was for the development of breeds for maximum yields of meat and milk, and which would attain a rapid and early maturity, a difference between chemistry and genetics. The Franco-Prussian war of 1870–71 removed Bechelbronn from French influence but long before this date government interest was directed towards the *Institut national agronomique* at Versailles which, opening in 1850, closed in 1852 to be reopened in 1876.

Boussingault's renunciation of large-scale agricultural projects in 1848 may not have been a purposeful gesture but the result of circumstances beyond his control. In 1847 he moved from the LeBel residence at Bechelbronn to the former Capuchin monastery at Liebfrauenberg[35] where, at a distance of five kilometres he was less able to supervise farm operations. Liebfrauenberg was not a farm although it possessed an excellent vegetable garden, the soil of which was frequently referred to in his later work on the nitrification of the soil. The farm managerial power behind the experiments from 1836 to 1848 had been exercised by his brother-in-law, Frédéric Achille LeBel, and latterly by Eugène Oppermann, a local farmer, which suggests that LeBel was spending more time on his petroleum interests. It was most likely that Boussingault's winter post at the *Conservatoire des Arts et Métiers* from 1845 onwards prevented the renewal of large scale farming experiments. Whatever the reason, although from 1836 to 1848 Boussingault was a field agronomist, yet after 1851 he was mainly a laboratory chemist concerned with the question of soil nitrification.

CHAPTER 8

CONSERVATOIRE DES ARTS ET METIERS

On the right bank of the Seine the wide and busy boulevard de Sebastopol proceeds northwards and near where it crosses the rue Réaumer is the rue Saint-Martin in which is found the *Conservatoire des Arts et Métiers*, both a technical college and a science museum. The facade of the Conservatoire which we see today was constructed in 1845, the year in which Boussingault was appointed to one of the two chairs of agriculture. Behind the Conservatoire is the rue Vaucanson which commemorates the man whose work formed the nucleus from which grew this institution.

Jacques de Vaucanson was an engineer who designed the machinery for supplying water to Lyon, and his studies of mechanics, music and anatomy enabled him to construct working models of musicians playing their instruments, and such was his skill that for the production of a play on the life of Cleopatra he fashioned an asp which not only writhed and hissed but at the appropriate moment struck at the bosom of the unfortunate queen. But apart from these ingenious novelties he also constructed working models of agricultural and industrial machinery, all of which were stored at his seventeenth-century Mansart designed house, the Hôtel de Mortagne, in the rue de Charonne. His ability was recognised by the *Académie des Sciences* of which he was made a member in 1748 and after his death a suitable place for the models was sought. A Convention decree of 1794 intended them to be exhibited in the newly established Conservatoire at the old abbey of Saint-Martin-des-Champs in the rue Saint-Martin, but the abbey was not free for occupation until 1798 as it had been an arms factory for the Revolutionary armies. Meanwhile, a school of geometrical drawing together with a technical library and workshop was established temporarily in 1796 at the Hôtel de Mortagne until the tenants of the abbey could be evicted, and it was not until 1802 that the Conservatoire was opened in the abbey by the chemist-statesman Jean Antoine Chaptal who was then Minister of the Interior.

Chaptal created not only the Conservatoire but chambers of commerce and a network of roads and canals as well as establishing the first sulphuric acid plant in France. The models and machines required a mechanic for their maintenance and the first to be employed was Joseph Marie Jacquard who perfected the loom so that it could be manipulated by one worker only instead of several which were previously required. In 1819 a *lycée* was started for the application of science to commerce and industry, receiving university status in 1843. At the time of Boussingault's appointment there were chairs in physics, descriptive geometry, applied geometry, mechanics, applied chemistry, industrial chemistry – the difference between the last two is not explained – agriculture, industrial economy, industrial legislation and ceramic arts.[1] One of the two chairs of agriculture was held by Louis Moll who taught farm mechanics and farm administration. The second chair, concerned with the teaching of crop and animal husbandry, was held before 1845 by Oscar LeClerc-Thouin who is little known today except for his theories of crop rotation in which the choice of crops in a rotation was governed by the depredations of insect pests or the ability to control weeds.

Boussingault Is Appointed to the Conservatoire

At a meeting of the Conservatoire Council, the *Conseil de Perfectionnement*, in January 1845 a letter was read from Mme LeClerc-Thouin announcing the death of her husband. The reaction of the Council was similar to any previous announcement of death or retirement when there was invariably an immediate demand that the syllabus should be revised or that the chair should be abolished and another established in its place in a subject sometimes remote from the original.[2] The Conservatoire was the responsibility not of the Minister of Public Instruction as might be expected, but of the Minister of Commerce and Industry, and he was not averse to suggesting new courses as on the occasion when he proposed a course in tropical agriculture! With true professional aplomb, the Council assured the Minister of their deep interest in this subject but that such a course was unnecessary.[3] At the Council meeting of January 24th, 1845 it was announced that there were six candidates for the vacant chair[4] of whom only Boussingault and Dezeimeris were seriously considered. Jean Dezeimeris was librarian of the faculty of Medicine in Paris, and his claim to be considered for a chair of agriculture was supported by his ownership of a farm in the Dordogne which was managed by a servant, distinguished from other servants by his ability to read and write. Farming instructions were sent from Paris by post![5] In support of Boussingault's application there was an impressive list of original papers from 1834 onwards and it was clear that he relied not on his South American experiences but on those gained at Bechelbronn. The Council was in favour of appointing Boussingault[6] but it was essential that the successful candidate should have the approval of the *Académie des Sciences* which was at once given, but a ballot became necessary a week later because someone had proposed Dezeimeris, nevertheless Boussingault triumphed with forty-three votes compared with only one for his rival.[7]

Before submitting Boussingault's election to the king for royal approval, the Minister wrote to the Council requesting advice from the members on the designation given to each of the agricultural courses. The president of the Council, Col. Arthur Morin, had asked Boussingault to attend a meeting to discuss the Minister's letter but Boussingault was unable to attend as his presence was required at Pechelbronn where a petroleum vapour explosion, killing five workers, had occurred on June 16th, 1845 at the Madelaine mine, due to a faulty safety lamp.[8] Previously the safety lamps had been tested in a mixture of hydrogen and air, but Boussingault thereupon devised a more certain test using ether at the bottom of a cylindrical tin into which the burning lamp was placed. If the ether vapour burned outside the lamp then the lamp was considered to be faulty, as a result of which the local Préfet ordered that all safety lamps were to be tested in this manner each week.

To return to the agricultural courses – they had been named *Cours de culture* and *Mécanique agricole* but the Council then decided that each was to be named *Cours d'agriculture* but that the title should be followed by a brief description, which for Boussingault's course reads:

> M. Boussingault's course will contain in general the following sections – climate, nature of the soil, fertility and manures, agricultural production considered with regard to the improvement of the soil and stock feeding, the comparative cultivation of cereals, fodder crops and cash crops.[9]

A letter of 27th December, 1845 from the Minister named Boussingault as the new professor according to an ordinance of the king, Louis-Philippe. Little is heard of Boussingault wining and dining but that day he invited M. and Mme Dumas to dinner at rue du Pas de la Mule with a further invitation to Dumas to give his opinion of the quality of a pâté de foie gras prepared with liver fattened by a new method contrary to their theoretical ideas.[10] It was probably prepared from liver supplied by Persoz but the letter suggests that Dumas was not yet a convert to the belief that fat was produced by a conversion of carbohydrate. The letter also informs us that the Boussingault ménage was no longer at rue Mézières which was given up around 1840 for a residence on the north bank at No. 8 rue du Pas de la Mule which leads off the place des Vosges in the Marais, that gem of early seventeenth-century architecture which we owe to Henry IV. The Conservatoire appointment must have meant some consolation to Boussingault in the year of the tragic mine explosion at Pechelbronn.

The Potato Blight of 1845

The justification for two chairs of agriculture was dramatically illustrated by the appearance of a potato disease known as potato blight, late blight or Peronospora in 1845 due to a parasitic fungus Phytophthora. Although the identification and prevention of fungus diseases are topics within the sphere of plant pathology, this subject was as yet unknown at the Conservatoire, and it seemed that plant diseases could be best studied by research in agricultural chemistry which the Minister wished to develop to combat this disease.[11] Not only were funds promised but land for experiments was reserved at the *Institut agronomique de Grignon*, sixteen kilometres from Versailles, which had a reputation second to none in Europe. It is difficult to believe that an advanced institution such as the Conservatoire, possessing two chairs of agriculture, was devoid of any land on which to practice its agricultural principles. The Grignon offer represented a major policy change. A committee was formed consisting of the professors of chemistry and agriculture, who proposed a research programme as follows:

(i) All necessary experiments should be undertaken to show the positive characteristics of the potato disease, whether it is new or whether it has occurred before.

(ii) Experiments should be undertaken on the means of preserving or making use of the crop.

(iii) Investigations should be made on the prevention of the disease by seed treatment, exposure to air and light of healthy tubers and those partly attacked.

(iv) Investigations should be made to determine whether any differences can be detected between early and late varieties of potatoes, light and heavy soils, dry and humid soils; and is the disease transmitted by soil from which diseased potatoes have been taken?

(v) Trials should be made in which the disease is induced.

(vi) Investigations should be made of the effect of mineral salts and organic substances.

(vii) Is the disease the result of storage?

(viii) Were there other cereals which could replace potatoes? It had been observed that when the potato crop failed the wheat crop was also poor but that the oat crop was good.

Following this business-like programme, reports were made based on experiments at Grignon which were summarised by Boussingault in his *Economie rurale*. Although the committee did not find a specific cure for the disease it identified the red granular matter in affected tubers as a fungus and almost as much general information was obtained as is known today. The committee recommended the use of fungicides such as ash or slaked lime, but alum, although effective, was not recommended because of its cost. The search for disease-resistant varieties was pursued, especially at Bechelbronn, but Boussingault was sceptical about imported varieties with a claim to immunity because European climatic conditions provided the ideal environment for the transmission of the disease, and therefore he recommended late sowing to avoid the cold and damp conditions that favoured it.

Boussingault's Syllabuses and Lectures – Appendices 1 and 2

At the end of each year the professors' syllabuses for the following year were scrutinised and criticised at a meeting of the Council, and it is for this reason that something is known of Boussingault's course in agriculture which appears to be an extraordinary mixture of the elementary and the complex, from the simple analysis of air using phosphorus to the estimation of glucose in sugar beet. Even the theory lectures contain such elementary material as 'Organic substances in the atmosphere' followed in a few weeks by 'Analytical control of the products of an oil works', owing its inclusion to the oil industry at Pechelbronn. Other influences can be traced, such as South America, the composition of the air, nitrification and animal nutrition, and in general the syllabus represented his many-sided experience and his multitudinous research interests as well as the influence of the abbé Branthôme at Strasbourg in 1821. If a fundamental weakness can be discerned, apart from frequent repetition, it is that he was teaching a course to comparatively elementary students which included much recent research which had not as yet been sufficiently verified by others. The syllabuses suggest a suitable course for research agriculturists desirous of knowing the advances taking place in fields other than their own specialities.

The standard and content of his lectures can only be judged by one of them, on the subject of the manure pit, which was published in 1858 as *La fosse à fumier*[12] which gives an indication not only of the standard of the lectures but also the level of comprehension of the students. Briefly, the lecture started with the need to conserve excrement and urine on a farm, with some practical details of the methods of transporting them to the fields with the minimum of loss. This was followed by a lucid, yet simple description of digestive processes, the composition of stock rations and the function of each constituent.

An interesting section of the lecture was the estimation of manure expected from a given quantity of fodder when fed to fully grown stock, a method due to Thaer in which the hay equivalents of the constituents of the fodder were added and the sum doubled to find the weight of manure expected. If 153 kilos of oats is equivalent to 241 kilos of hay, then this amount of oats fed to stock will yield 482 kilos of manure, consisting of wet stools and urine. Boussingault suggested that the hay equivalent of the ration should not be doubled but multiplied by 2.13, a degree of accuracy which could not be justified by the nature of the materials and is one of several instances of his lack of concern for approximations. Such emphasis on manure production was a reflection of its importance in an era when artificial fertilisers were rarely used.

He completed the lecture with a detailed description of the construction of a manure pit for stables accommodating twenty-five beasts, of which he had practical experience in 1848 when he inherited a part of Bechelbronn but not the farm buildings. *La fosse à fumier* consisted of two manure pits and an underground tank to receive urine from the stables which was sprayed over the manure by a pump to induce fermentation, the manure then being stored for periods up to five months. Nowhere in the lecture are the farmer-students confused by theories and one looks in vain for references to debatable topics such as the function of humus and mineral salts in the nutrition of plants, or of evidence of ammonia in the atmosphere, all of which were subjects of discussion among agronomists. Instead, a distinctly practical approach is noted which belies the hint of advanced theory which is revealed in his syllabuses. The testimony of his demonstrators and students was that his teaching was lucid, enthusiastic and enlivened by anecdotes from his very full life.[13] His command of language appears to have been a striking feature of his lectures and this is borne out by the funeral oration which he delivered in 1846 when the botanist Henri Dutrochet was buried. The standard of spoken French among scientists is well illustrated in the funeral orations for fellow members of the Académie des Sciences. They are not occasions for analytical criticism of the deceased's achievements but a time for eulogy, and on reading these *funerailles* we may learn little about the deceased but much about the standard of oratory of the speaker.

Agricultural education, like any other, includes the examining of students and the award of certificates. At the Conservatoire they were unknown until 1900; this disregard of qualifications is amply illustrated by the following story told by the surveyor Col. Aimé Laussedat who succeeded Arthur Morin as Administrator of the Conservatoire in 1881. Louis Pasteur, the pioneer of biochemistry, was Administrator of the *Ecole Normale Supérieure* when there was an arrangement by which his students attended Boussingault's later course in Agricultural Chemistry, and Pasteur attended with them. Speaking to Laussedat, Boussingault said, "We do not turn out students, we have an audience. Can you see me," added Boussingault, smiling, "can you see me being told to examine Pasteur after he has come to hear me in order to hand him a diploma?"[14] Boussingault was unfairly making an exception in an attempt to justify a general rule. This refusal to admit that all students were not Pasteurs resulted in a falling off in numbers which was especially marked after 1890 when similar institutions were attracting larger enrolments through their granting of certificates after suitable examinations. Perhaps he was influenced by memories of the public courses which he attended in Paris as a youth; instead of university degree courses he was conducting what would nowadays be recognised as a university extension course.

The Move to Liebfrauenberg

Letters written after Boussingault's return to Europe in 1832 were mainly addressed to Dumas, including one in August 1846 on the subject of animal nutrition which was sent from Bechelbronn.[15] However, in July 1847 he was writing to the astronomer, François Arago, from Liebfrauenberg,[16] enclosing a record of meteorological data made in 1846 by the curé of the neighbouring village of Goersdorf, M. Muller, with a request to supply the curé with a hygrometer to measure the humidity of the air which was usually high in the afforested regions of northern Alsace. The move to Liebfrauenberg was inevitable as the château LeBel at Bechelbronn belonged to the now expanding LeBel family. The fourteenth-century monastery of Liebfrauenberg had been purchased in 1825 by Boussingault's father-in-law, Marie Joseph LeBel and bequeathed to him before his death in 1842. Both Marie Joseph

Fig. 15. The Boussingault home in Alsace, the former monastery of Liebfrauenberg with one of the domestic staff in the foreground. (Collection: M. Charles Boussingault.)

and his wife were buried in the chapel of the monastery but their remains were transferred to nearby Lampertsloch cemetery in 1907. Boussingault installed a laboratory in the apse of the church, a partition separating it from the nave which became a vast store. The walls of the monks' cells were removed to provide a library and a lounge.

Liebfrauenberg has an interesting history. About the year 1350, a statue of Our Lady was discovered in a hollow oak tree by a shepherd walking on the hill which came to be known as Liebefrauenberg. The discovery started an annual pilgrimage to 'Our Lady of the Oak' which continued until the Revolution. A chapel was erected on the hill in 1384 and this date is inscribed on the lower part of the existing tower. In 1518 a Count Rheinard de Zweibrucken, having been cured of paralysis on one such pilgrimage, removed the crumbling nave of the old chapel and replaced it with three naves. As a result of the Reformation, the neighbouring village of Goersdorf became Protestant in 1571 whereupon the owner of the chapel demolished the naves and removed the stone to Morsbronn, seven kilometres asay, to build a Catholic church. Meanwhile 'Our Lady of the Oak' found refuge in the monastery of nearby Biblesheim, yet the pilgrimage continued, which is not at all surprising as her loyalty to the district had landed her in various uncomfortable situations. She survived a fire, a burial, and was later recovered, intact fortunately, from a local stream and eventually found a suitable place for retirement in the Catholic church at Goersdorf, built in 1900, whence there is a grand pilgrimage each Whit-Monday to the summit of the Liebfrauenberg. In 1717, Capuchins from the Alsace town of Hagenhau reconstructed the monastery but were compelled to leave during the Revolution, after which the estate was auctioned, there being several owners before it came into the possession of the LeBel family.[17]

Boussingault was now part of a feudalism which had persisted into the nineteenth century, for he had his town house, justified by his chair at the Conservatoire, and his country house. Besides Bechelbronn and Liebfrauenberg the LeBel-Boussingault families owned a small vineyard, Schmalsberg, and two other farms, Schwabwiller and Durrenbach, which were sold in 1859. Yet eighty per cent of local farmers possessed less than three hectares, the largest farms belonging to district councils and those of high social standing. The small farmers grew their own food, kept chickens and hares, and each farmer usually had four cows and two pigs. The farmhouse consisted of ground floor only, which sheltered the family and the beasts, a type of building which can be seen today in the northern Vosges. All clothes were made at home except shoes which were for adults only and worn on Sundays

Fig. 16. Each Whit-Monday there is a grand pilgrimage from the Catholic church in Goersdorf to the summit of the Liebfrauenberg. In the foreground is the former pilgrims' hostel, Greyenbuhl. (Photograph: F.W.J. McC.)

Fig. 17. The Pilgrims' Steps leading to the 14th-century tower of Liebfrauenberg. (Photograph: F.W.J. McC.)

to Mass or Vespers. Clogs were also purchased but were made locally. Although the farmers sold a little of their produce such as colza oil for lamps, thread and woven material, it was a closed economy, marked by the diligence and thrift so characteristic of peasant family organisation.[18] Cooperation within the family was general throughout Alsace and as an example there is the testimony of the engineer of the Pechelbronn oilfield who recalled in 1918 a conversation with Alice Crozet, Boussingault's younger daughter, in which she said:

> . . . that it was a tradition of the LeBels and the Boussingaults that each one, young and old, had to do their best for the head of the family in whatever work was being done. Thus, her brother had to supervise his father's laboratory distillations instead of running about the fields as a small boy of his age would have preferred. Her share was to take part in Boussingault's experiments on food assimilation by animals. As a very small girl she had started the work of looking after the hens whilst taking their temperatures, but sometimes to her great despair they escaped from her small hands and took refuge in the farm courtyard with a long thermometer fixed in the right place.[19]

Thus Boussingault in 1846 started on one of the most unusual and unorthodox of professorships, delivering two one-hour lectures per week in a session that lasted from November to April,[20] and spending the rest of the year on his farm in Alsace, living with his family in the spacious former monastery of Liebfrauenberg, now equipped with library and laboratory. The students were not lycée products anxious to obtain an agricultural qualification but were farmers and farm-owners who ceased work during the winter and attended lectures at the Conservatoire.[21] There was no college farm, but Bechelbronn could be regarded as the field laboratory, kept at Boussingault's expense, or perhaps supported by the revenues from the sale of Pechelbronn oil, for the financial allocations to his department did not suggest that his farm was ever subsidised by the Conservatoire.

CHAPTER 9

THE CONSERVATOIRE AND POLITICS

Politically, Boussingault tended to be liberal or left of centre which caused not a little anxiety at home in 1820, when there were student riots in Paris.[1] We have seen that trade depression provoked in him reactionary thoughts when blame was placed on parliament, press freedom and individual liberty. Yet his liberalism was reinforced by his association with Simon Bolivar to whom he gave moral support for the liberation of South America from Spanish domination. Boussingault's brief carrer as a professional politician from 1848 to 1851 started with a non-political incident which had far-reaching professional and political repercussions. He quarrelled with his young demonstrator at the Conservatoire, Georges Ville, who had been appointed in January 1848 and resigned protestingly in March 1849.[2]

Who was this Georges Ville? When Napoleon Bonaparte apportioned the conquered countries of Europe to his brothers, Louis Bonaparte received in 1806 the newly established throne of Holland. Louis Bonaparte was married, though rather indifferently, to Hortense Beauharnais, the daughter of Josephine, Napoleon's first Empress, and her previous husband, the Vicomte de Beauharnais. Louis was not interested in kingship; he had little sympathy with his brother's efforts to form a European bloc and made the cession of Brabant to France an excuse to abdicate in 1810. Louis found greater happiness with the housemaids of Hortense and to one of these was born Georges Ville at Port Saint-Esprit (Gard), taking his name from Georges Ville, a police commissioner at Lyon who was asked to marry the housemaid.[3] Hortense, just as unfaithful, had a lover, the comte de Flahaut, as a result of which Charles, the future duc de Morny, was born in 1811. Ville and Morny were, therefore, the half-brothers of Louis Napoleon Bonaparte, a legitimate son of Louis and Hortense, who was to become Napoleon III in 1852 following a *coup d'état* in 1851 which was organised by Morny. Little is known of Ville's early life except that he was one of Boussingault's students at the Conservatoire,[4] probably in 1846, and that he left for a post at the *Ecole de Pharmacie* before returning to the Conservatoire in 1848 as demonstrator to Boussingault.

Prior to Ville's resignation from the Conservatoire, there had been symptoms of mutual incompatibility which began when Ville, then on the staff

Fig. 18. Georges Ville (Bib. cent. Muséum d'Histoire Naturelle).

of the *Ecole de Pharmacie*, had given up fifteen days to analyses requested by Boussingault, yet when appointed to the Conservatoire, Ville received only one-third of the promised salary. This was one of many complaints which he submitted in a letter of resignation which ran to about fifteen hundred words.[5] It was also tactless and unprofessional for Boussingault to remind Ville in his letter of appointment that his duties were to be carried out in a conscientious manner; his reliance on Ville's analytical skill should have rendered such a statement unnecessary. As Boussingault now spent half the year at his newly acquired Alsace estate, it seemed reasonable to suppose that Ville was sometimes compelled to use his own discretion and judgement in making necessary purchases of laboratory materials, yet he was accnsed of overspending and purchasing goods without Boussingault's prior authority. In asking that materials for Ville's private research should be placed on his own private account, it suggested that he was making illegal use of departmental funds. Further complaints from Boussingault were that small quantities of reagents should be prepared as required, a result, perhaps, of his austere upbringing.

Ville was not the man to accept quietly these accusations of overspending and misappropriation of departmental funds, especially when he had already discovered what he supposed was a scandal. "I learnt yesterday from M. Boyeau, the chemical supplier to the Conservatoire, that M. Payen had given him one hundred francs from funds allowed us". It was the custom of the Conservatoire to make advances to the contractors, but Payen and Boussingault had separate allocations of finance. There was, perhaps, a mutual arrangement between them because Payen had collaborated with Boussingault in 1840–41 in finding the comparative values of manures and fertilisers in return for which Boussingault assisted Payen in his researches on the composition of foodstuffs. Ville also reminded Boussingault that when their laboratory assistant left in March 1848, they employed at various times labourers unknown to the Conservatoire staff. The regular Conservatoire labourers were only too familiar to the staff, for among the six cleaners there had been only four vacancies in twenty years.[6] The foodstuffs requisitioned for the Conservatoire had been too much of a temptation for these strange labourers, but as the accountants would have observed a re-ordering of the stolen items, Boussingault arranged with the grocer, M. Connelier, to include a list of other foodstuffs which were paid for but never delivered, or as Ville described them in his long letter, "objets de consommation fictifs". Ville proved most conclusively that he had paid for his own reagents from his private account.

Fig. 19. Liebfrauenberg: the laboratory (above) and library (below). (Photographs from *J. B. Boussingault à la Société Centrale d'Agriculture*, by M. Lenglen (Imprimerie Centrale Administrative, Beauvais.)

Having defended himself, Ville then attacked Boussingault in his letter; in addition to the broken promise concerning salary, Ville complained bitterly that he had been called upon to perform 110 analyses unconnected with the agricultural courses at the Conservatoire, probably part of Boussingault's animal nutrition experiments which were being conducted at that time. He was also accused of lack of interest in Ville's experiments: "Not ten days ago at the Conservatoire, a few paces from my apparatus, I asked you to come and see it and you again refused". It was the letter of a paranoic only too willing to assume the rôle of martyr. Undoubtedly Ville had been treated shabbily by his professor, yet he could be sufficiently magnanimous to end his letter on a friendly note, "But I shall try to forget these troubles and recall only the kind welcome you have accorded me in the past and, above all, the encouraging goodwill with which Mme Boussingault has always favoured me". At this time Boussingault could not have realised the consequences of this incident which led to bitter clashes between the two men on several occasions, and proabbly prevented him from becoming a peer of France, an honour which he merited for his pioneer work in agricultural chemistry. Other scientists had succeeded, Chaptal, Dumas and his friend Henri Saint-Claire-Deville, who is known for discovering a method of extracting aluminium from its ores, and perhaps Boussingault wished to succeed for he wrote to Dumas in 1872, "I have spent my life as I always spend it – I am only a mister but I attract all the little children of the Peerage".[7]

Boussingault Is Elected to the Assemblée Nationale

Louis Philippe's government under Guizot, who was now Prime Minister, was a government of the centre and therefore vulnerable to attacks from the Right and the Left. The Right consisted of clericalists who had constantly opposed the growth of secular education when Guizot was Minister of Public Instruction. It also included legitimists who supported the claim to succession of Henri de Chambord, Duc de Bordeaux, the son of the assassinated Duc de Berry, heir to Charles X. The Left had been out of favour since the 1834 revolts in Paris and Lyon, and the attempted assassination of Louis-Philippe in 1835. Differences among the various sections of the Left were forgotten when they united under the principle of universal manhood suffrage. The bad harvest of 1846 and growing unemployment due to an industrial slump were conducive to the hasty action which Louis-Philippe took in dismissing his now unpopular Guizot. Only an incident was needed to spark off a revolution and this was provided by his own troops when a peaceful demonstration outside the Ministry of Foreign Affairs was pushed forwards by the crowds behind. Panicking, the troops fired and the resulting deaths incensed the Paris mob to take control of the city. Louis-Philippe, realising the hopelessness of the situation, did not call on the army to help but abdicated on February 24th, 1848 in favour of his ten-year-old grandson, Philippe, comte de Paris, but the people did not relish another monarchy and thus the Second Republic came into being, administered by a provisional government. It introduced universal manhood suffrage with the result that the electorate rose from a quarter of a million to nine millions.

An election for the *Assemblée Nationale* was decreed for April 1848, in which year there had been formed a workers' party, the *Société centrale des ouvriers*, a party well left of centre formed by Louis Blanc. The party's philosophy had been set forth in Blanc's book, *l'Organisation du travail*, published in 1840, and briefly expressed by the slogan, "Let each produce according to his aptitudes and his strength; let each consume according to his needs", more concisely stated later by Karl Marx, "From each according to his abilities, to each according to his needs".[8] Blanc was against competition and the artificial value of products; he wanted government control of industry and the prohibition of children in factories.

In Alsace the local branch of Blanc's party, the *Comité centrale de Strasbourg*, held a meeting on March 19th, 1848 to receive the names of their candidates as recommended by the various cantons of the Bas-Rhin département. Boussingault's name was presented by several cantons and finally included in a list of fifteen, which was the allocation for the Bas-Rhin. In each canton there were about 70 candidates but some were not so confident about Boussingault, such as Lauterberg on the Rhine which placed him 20th in the poll, but he fared better elsewhere so that finally he came twelfth in the total poll.[9] In supplementary elections in June 1848, Louis Napoleon Bonaparte was elected and as step-brother to Ville this event was to bring tribulation to Boussingault. It is not clear why he stood as a candidate for Blanc's party except that it appeared to be the best organised party in Alsace. In the *Assemblée* his bourgeois sympathies clashed with the views of his party and henceforth he aligned himself with the moderate Republicans. He even voted in favour of the arrest of his leader, Louis Blanc, who was involved in an insurrection at the Palais Bourbon where the *Assemblée* met.[10] Boussingault also voted against the principle that there was a right to employment; and against a proposal to abolish the death sentence. Yet he opposed attempts to concentrate power in the hands of one or few individuals such as the Grévy amendment, a well-intentioned proposal of Jules Grévy to place executive power with a cabinet president in order to forestall a possible seizure of power by Louis Napoleon Bonaparte.[11] The amendment might not have been defeated had the *Assemblée* foreseen the *coup d'état* of December 1851, by which Louis Napoleon eventually became the Emperor, Napolean III.

Boussingault Debates Agricultural Education and Becomes a Conseiller d'État

Two significant events in Boussingault's short parliamentary career were the debate on agricultural education in 1848[12] and his election to the *Conseil d'Etat* in 1849.[13] His sole contribution to debates at the *Assemblée* was a short intervention at the end of a protracted debate on agricultural education concerning the establishment of farm schools – one in every canton; a farm college in each département; and an institution of university rank to be known as the *Institut agronomique*. The representatives were in no doubt about the need for the *Institut agronomique* and passed this section of the bill with only token opposition. The other proposals provided much of the debate, especially as to who would pay for them – should the schools be supported by the taxpayer or should they be private schools? Was a farm school really needed or would a model farm prove more suitable? With incredible naiveté some representatives assumed that a model farm would automatically raise the standard of the surrounding farms by its example. Towards the end of the debate, the *Assemblée*, perhaps a little wearied of the subject and ready for any light or humorous diversion, Boussingault was inadvertently the cause of much merriment. Jean Dezeimeris, one of the candidates for the Conservatoire chair of agriculture in 1845, represented the Dordogne, and in an amendment he suggested that in each commune the state should offer a subsidy of five hundred francs to the farmer who had given up the greatest proportion of his land to fodder crops, the subsidy being limited to those who had at least six hectares under cultivation. Boussingault reminded him that, except for the eastern part of France, the average farm was of two to three hectares and the proposal was, therefore, undemocratic. But let Boussingault continue in his own words:

I know that the honourable M. Dezeimeris is a great believer in fodder crops (laughter). Each man to his own taste (boisterous laughter). Dear me, gentlemen, my thoughts have been misunderstood (from all parts, 'No! No! just

Fig. 20. Boussingault in 1849 as a Conseiller d'Etat. (Collection: Mme F. Boussingault.)

a silly joke! Carry on!'). A fodder crop is meat. I respect and even agree to a certain extent with the views of our honourable colleague (renewed laughter) but I cannot prevent myself from saying that it is difficult to admit that, everywhere and always, the excessive planting of fodder crops is the most advantageous or profitable farming. It seems that when it is a question of deciding on this or that agricultural system we should take into account the nature of the soil, the climate, and above all the commercial prospects. Furthermore, I believe that the Assemblée Nationale should beware of decreeing which agriculture is the simplest, the most productive and the most fruitful. I vote against the amendment ('Very good!').[14]

The bill became law, the *Institut agronomique* was established at Versailles but the opening of the colleges and farm schools proved a slow process.

The *Conseil d'Etat* was a council established by Napoleon in 1799 when he was First Consul, its function being that of a technical advisory board and as such it had great influence under the Consulate and the Empire, but had thereafter declined in power until the advent of Louis-Philippe. The members had been elected by the cabinet but after 1848 they were now elected by the *Assemblée.*[15] Boussingault's name appeared on a list of eighty-four candidates in March 1849 from which forty were to be selected by a ballot which had a built-in procrastinating device, which insisted that successful candidates should poll a figure at least equal to half the number of voters. In the first ballot Boussingault was 59th and fared little better in succeeding ballots. After seven ballots, representatives were complaining of the waste of time, and although 39 of the 40 members had been elected it required three more ballots to complete the membership of the *Conseil d'Etat*. Due to the withdrawal of a candidate Boussingault secured the necessary votes and just succeeded as the fortieth member.[16]

He served on the *Conseil d'Etat* from April 1849 to December 1851, and had the reputation of acquitting himself well.[17] His specific contributions cannot be traced because practically all the records of the Conseil were lost by fire in 1871, when the Communards, in fighting the army in Paris, burnt down public buildings to facilitate their fire power. The Palais d'Orsay which housed the library of the *Conseil d'Etat* was not spared; but a few slightly charred documents of the period 1849–51 survived the fire and tell of decisions of the *Conseil* on requests from farmers who sought permission to drain marshes, to construct dams, or from those who made claims for land lost due to the re-aligning of roads.

An election for the first president of the Second Republic was held in December 1848 and the small Bonapartist following proposed Louis Napolean Bonaparte who appeared to have no hope whatsoever of becoming the President. But there were many representatives who were disappointed with the results of the 1848 revolution which they themselves had engineered. Moreover, the peasantry feared a possible confiscation of property and thus looked for a political saviour. These two reasons, coupled with a publicity campaign and some emphasis on the Napoleonic legend created by his famous uncle, enabled Louis Napoleon Bonaparte to become the first President of the Second Republic. The architects of the new *Conseil d'Etat* had not foreseen this possibility, for they had planned it as a legislative and administrative body to act as a second chamber between the executive power of the President and the legislative power of the *Assemblée*. Previously it had been a council of the monarchy and it reverted to this status after Louis Napoleon Bonaparte's coup d'état of December 1851, followed by his assumption of kingship in January 1852 when, once again, it was the Emperror who nominated the members.[18]

An Attempt to Oust Boussingault from the Conservatoire

At a meeting of the Conservatoire Council – the *Conseil de Perfectionnement* – on November 3rd, 1849 a letter was read from the Minister of Agriculture and Commerce pointing out that Boussingault, as a salaried *Conseiller d'Etat*, could not receive another salary from the Conservatoire, and that therefore his resignation from the Conservatoire had been assumed from the day he had accepted the appointment of *Conseiller d'Etat*. The Administrator of the Conservatoire, Arthur Morin, was asked to implement the order, but after the Minister's letter had been read to the Conservatoire Council it was unanimously agreed to recommend that Boussingault be given the title of honorary professor. It was also agreed by the Council to discuss a successor to Boussingault at the next meeting but the decision was postponed because the end of the year was close at hand. There was a further complication because now that the professor had been assumed to have resigned, his demonstrator, Houzeau, could not be officially recognised and indeed had received no salary for the period April to November 1849. A new Minister of Agriculture and Commerce had been appointed – no other than Dumas, eminent scientist and very correct civil servant – who approved the claim for salary on behalf of Houzeau but asked from which fund was it to be paid. As if Dumas had some premonition, or even inner knowledge, of things to come, he wrote that he agreed to the proposal to make Boussingault an honorary professor but since the *Institut agronomique* at Versailles would have a chair of agriculture, a second chair at the Conservatoire would be unnecessary. Then Dumas added that he disagreed with the last statement and could hardly conceal his disapproval of the blow that was to fall.[19] Almost two years were to elapse before any further action was taken, during which attempts were made to change the law which forbade a President from continuing in office for a second term, a procedure which was approved by the *Assemblée* but not by the necessary two-thirds majority. Undeterred, Louis Napoleon and his friends decided to act without seeking the approval of the *assemblée* and on the night of December 1st–2nd, troops occupied the main strategic points in Paris where resistance was crushed by 30 000 troops. Following a favourable plebiscite, the Second Empire was declared in 1852 with Napoleon III as Emperor.

Boussingault did not attend the meeting of the Conservatoire Council on December 11th, 1851, nine days after the *coup d'état*. The Administrator of the Conservatoire, Morin, had received a letter from the Minister, the contents of which had probably been conveyed previously to Boussingault. Morin read the letter to an astonished Council, for the Minister had abolished Boussingault's chair of agriculture and had substituted a new chair of agricultural chemistry. Furthermore, he had appointed to the new chair, of all people, Georges Ville!

As well as Administrator of the Conservatoire and professor of mechanics, Morin was a soldier and therefore a man of action. He did not wait for the next meeting of the Council but sent a three foolscap-page letter of protest to the Minister reminding him that there had been n o reply to the Council's resolution to retain Boussingault as an honorary professor in his chair of agriculture. Moreover, the correct method of appointment was for Ville to first present himself before the Council and the *Académie des Sciences*, as Morin had already advised him. In the same letter he informed the Minister that although he appreciated that Ville had conducted experiments at his own expense he questioned the value of the results especially after consulting several celebrated chemists who had expressed no confidence in his work. Now, Morin continued, since the *Conseil d'Etat* had been dissolved it was only right that Boussingault should return to his former post for which there had never been a replacement. "You would not want the demonstrator to supersede the professor by taking advantage of a political issue", Morin wrote strongly.[20] Yet in spite of his close relationship to the principal characters of the drama, Louis Napoleon and Morny, Ville thought it prudent not to accept the post. At the Convervatoire Council meeting of January 8th, 1852 Morin informed the members that the Minister had issued a government notice naming Boussingault as the professor of agricultural chemistry in succession to Ville! – for which the Council tactfully expressed its gratitude to the Minister for the conciliatory manner in which the affair had been conducted.[21] Ville's action in not accepting had prevented an explosive situation, but did the Conservatoire know, as Ville knew, that he had been appointed to the newly created chair of agricultural chemistry at the *Institut agronomique* at Versailles?[22] The Archbishop of Paris had sent him a congratulatory letter dated 1st January, 1852.[23]

Why, then, did Boussingault retain his professorship on being elected to the *Conseil d'Etat*, knowing that it was contrary to regulations? The answer is contained in an inspection report on the teaching at the Conservatoire, in the form of a letter from Delambre, Chief of the Division of Commerce, to his Minister, Dumas, dated February 1850. It referred to the proposed resignation of Boussingault on joining the *Conseil d'Etat*, but said that he had retained the post because there was an attempt to reduce the two agricultural courses to one course in order to establish a chair of agriculture at the *Institut agronomique.*[24] The figure that emerges with the greatest honour from this shameful episode is Arthur Morin, the Administrator of the Conservatoire whose firm and fair handling of the institution's management, coupled with his high sense of loyalty to his staff, undoubtedly contributed greatly to the success of the Conservatoire during these hazardous years.

Oxygen from the Air

Parliamentary duties at the *Assemblée Nationale* and the *Conseil d'Etat* meant that the summers and autumns of 1849 to 1851 were spent in Paris instead of Alsace. This gave Boussingault the opportunity to investigate at the Conservatoire the practical consequences of the reaction occurring when barium oxide, or baryta, is heated in air to form a higher oxide, which can then be decomposed to liberate atmospheric oxygen. This reversible reaction had been discovered by the French chemists Gay-Lussac and Thenard in 1810,[25] but it was Boussingault who not only proved that the higher oxide was barium dioxide but conducted experiments which prepared the way for the industrial preparation of oxygen from the air. He realised that the reaction between baryta and the air could be made a repetitive cycle, by which an indefinite amount of oxygen could be obtained from the air. The results at first were far from encouraging; successive passages of air over the baryta yielded less and less oxygen. He did find, however, that instead of using dry air the moisture in undried air was sufficient to yield a much improved result. The method was partially successful but it never proceeded beyond the laboratory stage, where, as a source of oxygen, he kept a tube containing barium dioxide.[26]

Boussingault returned to this subject after twenty-seven years[27] and found the optimum conditions of temperature and pressure required to render the laboratory reaction a valid industrial venture from which he gained nothing. The Brin brothers patented a method in 1880 of obtaining oxygen from the air based on Boussingault's work of 1879[28] and their production of oxygen supplied the industrial market for twenty years, but eventually it was unable to face competition from methods using the liquefaction of air. Yet before the Brin brothers had started manufacturing industrial oxygen, Louis Cailletet of Paris and Raoul Pictet of Geneva had succeeded in liquefying atmospheric oxygen. There exists a roughly pencilled note, dated 24th December, 1877 from Pictet to Boussingault, "Oxygen liquefied today at 320 atmospheres and –140°C . . . " followed by a sketch of the apparatus.[29]

So the locust years of politics were not altogether wasted. He was, perhaps, too reserved to become a successful politician, and his quarrel with Ville became a quarrel with the Establishment. Boussingault's own conclusion on politics and science, "Few scientists have succeeded and science has always been the loser"[30] is only partially true. French political life has been noted for the contribution made by savants, either of the humanities or of the sciences.

CHAPTER 10

THE PLANT AND NITROGEN

Although throughout antiquity and medieval times there had accumulated much empirical knowledge on manures, fertilisers and the beneficial effect of legume crops, no theory of plant nutrition was possible until two facts had been established, namely, the discovery of nitrogen and the presence of nitrogen in plants. Henry Cavendish (1731–1810) distinguished 'mephitic air', or nitrogen, from 'fixed air' or carbon dioxide, and obtained nitrogen by the passage of air over heated charcoal. Joseph Priestley (1733–1804) distinguished between 'phlogisticated air', or nitrogen, and 'dephlogisticated air', or oxygen, and realised that the atmosphere consisted of two distinct parts. He was a phlogistonist, a follower of the theory that a burning substance gave off 'phlogiston' to the air. Carl Wilhelm Scheele (1742–1786) of Sweden claimed that the atmosphere consisted of two parts, 'vitiated air' and 'fire air'. But Daniel Rutherford (1749–1814) was the first to announce the discovery of nitrogen in 1772 using the method of Cavendish.

It was at the latter end of the eighteenth century that nitrogen was identified in plants and animals, first by Claude Louis Berthollet (1748–1822) in 1785,[1] and then by Antoine François Fourcroy (1755–1809) in 1789, whose discovery of nitrogen in the muscle of both carnivorous and herbivorous animals[2] suggested a connection between the nitrogen of plants and animals. That the vast reservoire of atmospheric nitrogen must contribute to the nitrogen of the plant was a rationalistic intuition first expressed by Jean Antoine Chaptal (1756–1832):

> The existence of nitrogen in certain products of vegetation is referable probably, in part, to a portion of this gas being introduced into plants by atmospheric air in combination with water which holds it in solution.[3]

Yet until about the middle of the nineteenth century the current theories of plant nutrition attributed the nitrogen of plant tissues either to the ammonia of the air, as stated by Justus von Liebig (1805–1873),[4] or to the nitrogen of manures as claimed by Sigismund Hermbstaedt (1773–1852) who was the first to use the term 'agricultural chemistry'. Boussingault was sceptical about both theories so he carried out pot trials in 1837 and 1838[5] in which oats, clover, wheat and peas were grown in sterile soil, formed by calcining sand,

and watered with distilled water. The expected gains in carbon, hydrogen and oxygen were obtained but a gain in nitrogen was recorded for clover and peas only and not for oats and wheat. Boussingault now had evidence that some plants, as he thought, could extract nitrogen from the air, and had he accepted these results a second confrontation between him and Ville, twenty years later, could have been avoided. Instead, Boussingault suspected that the clover and peas gained nitrogen because of the presence of ammonia and organic dust in the atmosphere, so he promised further investigations which were postponed to 1851 because of his animal nutrition trials and his political interlude.

Nitrogen Fixation Claimed by Ville

As a result of experiments in the Jardin des Carmes, Georges Ville claimed phenomenal increases in the nitrogen content of plants through assimilation of atmospheric nitrogen. His experiments of 1849–50 [6] consisted of growing plants in bell jars hermetically sealed at the base by standing each jar in a bowl of water. An atmosphere of air charged with carbon dioxide, to encourage growth, entered and left the bell jar, and so he claimed to have demonstrated nitrogen gains in rape, lupin, wheat, rye and maize. An expert on ammonia determination in air – he had measured it for each day for a whole year – he measured the amount of ammonia in the air entering and leaving the bell jars and concluded that there was no difference. Therefore he declared that the increase in nitrogen in the plant was at the expense of atmospheric nitrogen. So certain was he of his conclusion that he exclaimed, "Although the plants have not yet been gathered, one can consider the question settled". Actually, subsequent analyses proved that there had been increases in plant nitrogen.

Ville's findings assumed a greater importance and might have been generally accepted had not Boussingault fulfilled his earlier promise of 1838 to repeat his experiments of 1837 and 1838 using atmospheres free of ammonia and dust. His plants were treated to a continuous supply of carbon dioxide to promote growth but unlike Ville's plants, the atmosphere was not renewable. An analysis of the plants after two months growth showed that no nitrogen had been fixed.[7]

The opposite conclusions of Ville and Boussingault presaged another confrontation between the two men but who was to make the first move? Not unexpectedly, Ville defended his results before an open session of the

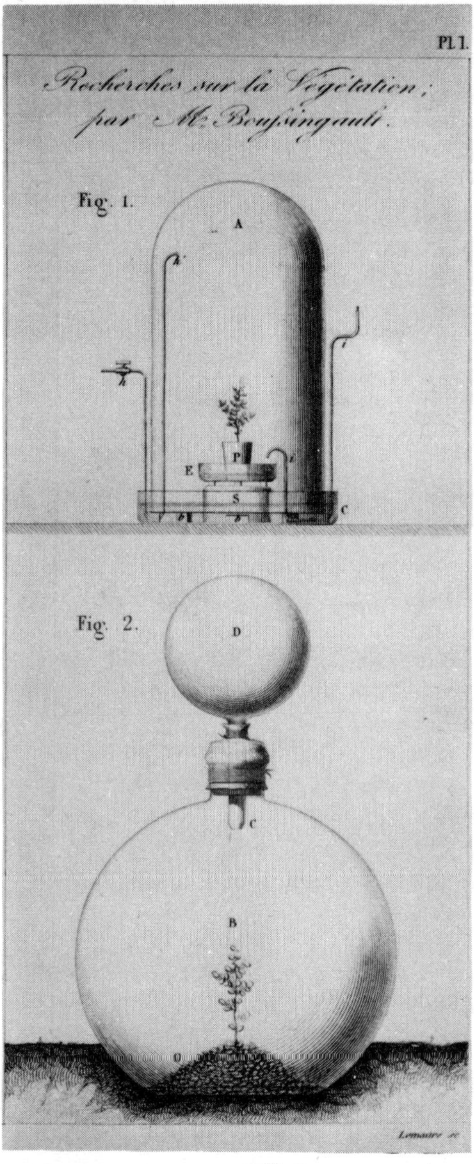

Fig. 21. Boussingault's experiments on nitrogen assimilation by plants in a non-renewable atmosphere. (From *Annales de chimie et de physique*, 1854, Vol. 41, pp. 5–60 (Masson).)

Académie des Sciences on April 10th, 1854 in which he criticised the nature of Boussingault's techniques;[8] he objected to his non-renewable atmospheres in closed flasks, and the growing of plants in porcelain rather than in earthenware crucibles, both conditions unnatural to plants. Ville was tempted to taunt Boussingault with having discovered nitrogen assimilation in 1837 and then denying it in 1853 to which Boussingault neatly replied, "I do not have to defend myself against an attack on my work; far from that, I have to defend myself from having made a discovery".[9] Boussingault promised further results from additional investigations and added. significantly and prophetically, ". . . in the conditions under which my plants have developed there is always a gain in nitrogen when the soil contains the most trifling quantity of nitrogenous matter . . .",[10] a remark which indicated a future trend in research on the nitrification of the soil.

The Conflicting Conclusions Are Examined

Because of these contrary conclusions Ville asked the *Académie des Sciences* to examine his experiments and he offered to repeat them before a committee [11] convened in 1854 under the chairmanship of the evergreen Michel Chevreul (1786–1889). The experiments, repeated outside a laboratory at the *Muséum d'Histoire Naturelle*, were fraught with difficulties and hazards from the start, when germinating seeds, supported on a newly painted lead sheet, died from the effects of turpentine fumes. During a second attempt the chemist in charge of the work was suddenly called away to Lille to the bedside of a dangerously ill father, and meanwhile some students in the nearby laboratory were separating iron and nickel salts, a technique requiring the use of ammonium hydroxide which is constantly giving off ammonia gas. Boussingault's reaction on hearing of this mishap was that it did not astonish him to learn that plants gained nitrogen when receiving water rich in ammonia![12]

The Committee drafted a masterly report but had certain criticisms of the task to which it had been called. Among them, it referred to the difficulties of conducting such work in the open air with variations of temperature, with hazardous winds and storms. It queried whether the same conclusions could be drawn from a crop several hundred times the weight of the seed as compared with a crop only several times the weight, a comparison between the results of Ville and Boussingault, or perhaps between commercial and laboratory practice. In spite of these and other criticisms, including a caution to

Ville for not revealing in 1849 that he had embarked on these experiments, the Committee concurred that the results of the experiments at the Muséum agreed with those previously made by Ville. Significantly, the Committee did not commit itself and declined to say that Ville had demonstrated the assimilation of nitrogen by plants. Hè did, however, acquire a reputation in some circles and some biographies, for having been the first to demonstrate this phenomenon, a reputation which has persisted erroneously to the present day. He was granted a sum of 2000 francs by the *Académie des Sciences* as a reimbursement and a further 2000 francs to assist in the continuation of the work.

That the Committee had been appointed did not please Boussingault who asked why Ville's experiment had been chosen for verification,[13] the answer, of course, was that Ville had requested it. This rebuff was interpreted as a lack of goodwill [14] and almost caused Boussingault to forsake Paris for the vacant rectorship of the University of Strasbourg.[15] Frequent cancellations and alterations in a letter on the subject to Dumas are those of a strained and nervous subject. He did not wish to be seen in Paris during the period of the Muséum experiment and obtained Ministerial approval to start his Conservatoire course at the end instead of at the middle of November.

The additional investigations promised by Boussingault were performed in an apparatus similar to that of Ville in which plants in a glass walled case were supplied with ammonia-free air and carbon diozide in a renewable atmosphere, yet he found no evidence of nitrogen fixation.[16] Repercussions of this controversy went beyond the walls of the Académie and the Muséum for it was a matter for private discussion among those who visited the Paris Exhibition of 1855 where Boussingault served on the International Jury of Agriculture.[17] At the time it was erroneously suggested that the rapid passage of air in Ville's apparatus induced a greater uptake of water containing dissolved atmospheric nitrogen.[18]

Viewing the controversy in retrospect, it is realised that Boussingault was right and Ville wrong, for there is no direct assimilation of atmospheric nitrogen by plants. The aspiration of large volumes of air at great speed in Ville's apparatus led to the then unknown soil bacteria falling on the soil of his flower pots. These, and not the plant, fixed atmospheric nitrogen which eventually reached the roots as nitrates, a process which was not demonstrated until after a lapse of thirty years. The discovery of atmospheric nitrogen fixation had been largely ignored since 1862, when F. V. Jodin exposed to the air a nitrogen-free solution of glycerol, tartaric acid and phosphates until a rich bacterial vegetation was obtained. Placed in a sealed

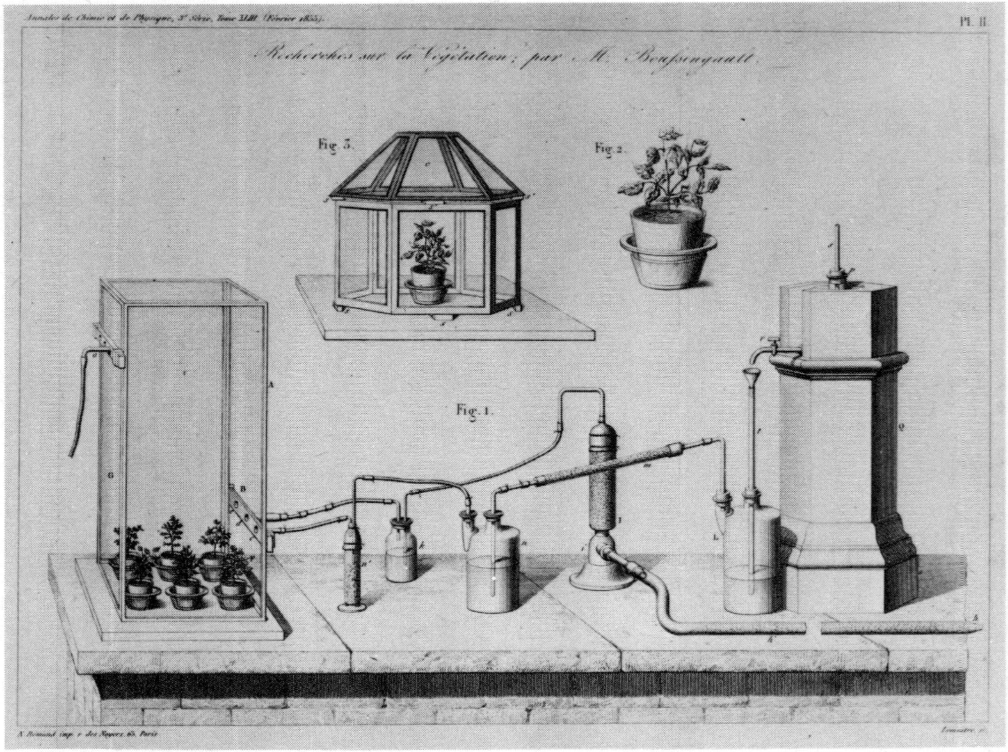

Fig. 22. Boussingault's experiments on nitrogen assimilation by plants in a renewable atmosphere. (From *Annales de chimie et de physique*, 1855, Vol. 43, pp. 149–223 (Masson).)

vessel containing nitrogen and oxygen there was a notable absorption of nitrogen.[19] It was not until 1893 that Serge Winogradsky (1856–1953) discovered nitrogen fixing bacteria in the soil,[20] and generally such bacteria become nitrogenous organic material converted by other bacteria first to ammonium compounds, then to nitrites and finally to nitrates which are able to function as plant nutrients.

The special case of the nitrogen fixing properties of legumes, discovered by Boussingault during his classical crop rotation experiments of 1836–41, was elucidated by H. Hellreigel and H. Wilfarth of Germany in 1888.[21] They made certain assumptions which were verified, namely, that nodules on the roots of legumes were induced by soil bacteria; that these bacteria fixed atmospheric nitrogen; and that legumes without nodules behaved like non-legumes in using nitrogen compounds in the soil rather than atmospheric nitrogen. Again, their discovery followed long after the discovery of bacteria in the nodules of legumes by J. Lachmann in 1858 who published his work in an obscure journal.[22]

Boussingault's negative findings were the result of eliminating contaminents such as ammonia and dust, and his insistence on sterile cultures which prevented the essential rôle of bacteria being realised. It is an example of the failure of an accurately controlled experiment which omits an important factor. Less accurate but frequently repeated trials in the field may yield a more precise answer. Ville repeated his experiments a year later,[23] but admitted to various disappointments for which he could not adduce reasons; no longer could he record the striking gains in plant nitrogen which had impressed the Chevreul Committee, but if there was a sense of failure in the accumulation of abortive attempts to establish the fact of atmospheric nitrogen fixation of plants, it was never shown in the writings of the two protagonists. Instead, they transferred their attention to the nature and function of nitrogenous plant nutrients in the soil. They were not the only workers in this field, as there was a plethora of experiments by others characterised by their artificiality, such as exposing the plant to ozone and ammonia which were supposed to react to form nitric acid; or growing a plant in an atmosphere of ammonium carbonate powder.

The Chevreul Committee experiments and those which followed proved an anti-climax, yet there were those who still felt that further investigation was desirable such as John Bennet Lawes, John Gilbert and a visiting colleague from the U.S.A., Evan Pugh, who published their Rothamsted results in 1861.[24] Their apparatus was similar to that of Ville and was undoubtedly the best organised investigation of its kind yet it gave no evidence of nitrogen fixation by plants and supplied no advance whatsoever on what was known from Boussingault's experiments of 1854. In spite of failures in their main and collateral experiments, the Rothamsted trio were clear in their minds as to the direction of future experiments:

In view of the evidence afforded of the non-assimilation of free nitrogen by plants under the wide range of circumstances provided in the experiments, it is desirable that the several actual or possible sources of *combined* nitrogen to plants should be more fully investigated both qualitatively and quantitatively.[25]

But long before the Rothamsted experiments such a start had been made with investigations by Boussingault, Ville and Frédéric Kuhlmann of Lille. They transferred their attention from the plant foliage to the soil and roots. The work of Boussingault, Ville, Lawes, Gilbert and Pugh had been hindered by two erroneous concepts, namely, the alleged nitrifying rôle of ammonia in the air, and the false analogy between the intake of nitrogen and carbon dioxide by leaves.

There appear to be no records of investigations similar to those of Boussingault, Ville and the Rothamsted workers due to an experimental dichotomy in plant nutrition research where emphasis was either on the contribution of the atmosphere or on the rôle of mineral salts. Whereas they were concerned with the possible function of atmospheric nitrogen, other workers in Britain and Germany concentrated on plant ash and soil analyses which, they believed, would indicate the source of plant elements as well as the nutritional demands of the plant. Sir Humphrey Davy (1778–1829) had already, in 1815, drafted a scheme for the analysis of soils,[26] and it is not surprising that one of the outstanding British agricultural chemists, Charles Giles Daubeny (1795–1867), preferred to concentrate on the analysis of soils and plant ash as this was the approach of his former mentor at Giessesn, Justus von Liebig. Yet Daubeny did demonstrate that the plant could only extract a small fraction of the nutrients of the soil and referred to the 'active' portion which could be extracted by a solution of carbonic acid which did not dissolve the 'dormant' fraction.[27] Lawes, who was a student of Daubeny at Oxford, and Gilbert, opposed the ash analysis approach, arguing that a small proportion of an element in plant ash might indicate that there was difficulty in extracting that element from the soil. They also dismissed the contention of Liebig and Daubeny that the source of plant nitrogen was the ammonia of the air, having found that it did not account for the nitrogen increase in a crop.[28]

Plant Nutrients in the Soil: The Rôle of Humus

As the attempts to confirm the assimilation of atmospheric nitrogen by plants had reached no definite conclusions, attention was directed to the soil. It appears that workers in this field realised subconsciously that the problem was divisible into two main parts, the accumulation of nitrogen in the soil, and its passage to the plant. It was the second of these that first occupied the attention of agricultural chemists, for it was difficult to visualise a solution of the problem of nitrogen accumulation in the soil without a knowledge of plant nutrition because a part of this nitrogen accumulation was of necessity composed of plant nutrient material without which plant development was impossible.

Until about the middle of the nineteenth century it was generally believed that plant food was obtained from humus, the dark material in the top layer of soil from which nutrients, extracted by water, were conveyed to the plant. This process especially applied, so it was believed, to the plant's source of carbon which is the main constitutent of humus. A host of soluble compounds was thought to be obtained from humus, and their names, now largely forgotten, are listed in an entertaining volume of the time, *A Muck Manual for Farmers*, by Samuel L. Dana and published in 1842. The standard of organic analysis of the day could not possibly cope with such a plethora of doubtful compounds and their complex formulae. The humus theory persisted during the first half of the nineteenth century in spite of the discovery of carbon dioxide assimilation, or photosynthesis as it is now known, at the beginning of that century, by which atmospheric carbon dioxide passes into the green parts of a plant by way of holes, or stomata, and forms carbohydrates in the presence of light and chlorophyll. Oxygen is expelled from the plant, a phenomenon first noticed by Joseph Priestley about 1775. Yet one of the outstanding discoverers of carbon dioxide assimilation, Nicholas Theodore de Saussure (1767–1845) of Geneva, was also a supporter of the humus theory. The mid-century witnessed a polemical exchange of views between the protagonists of the humus theory and those of the mineral theory of plant nutrition. Humus as the main plant nutrient, if not the sole source, had received important support from Jons Jacob Berzelius[29] (1779–1848) in his rôle of leader of chemical thought, and from Sir Humphrey Davy.[30] Gerard J. Mulder (1802–1880), a student of Berzelius, described a theory of plant nutrition in which humus yielded various acids, humic, ulmic, geic, etc. which formed soluble salts with ammonia, potash or soda, and thus entered the plant roots in solution.[31] The humus theory represented the culmination of a period of more scientific but untested beliefs which were able to flourish in the absence of an experimental theory of plant nutrition. In addition, the four ancient elements still influenced agricultural theory; air and water were now obviously essential; fire with its connection with saltpetre was admitted; there remained the earth with what seemed a store of nutriment in the form of humus.

The Mineral Theory

The mineral theory of nutrition is generally attributed to Justus von Liebig through his very famous treatise of 1840, *Organic Chemistry in its Applications to Agriculture and Physiology*, yet his ideas had been anticipated in 1838 by the German agriculturist Karl Sprengel (1787–1859)[32] who was convinced not only that plants could dispense with humus but that mineral salts in the soil were of more importance in plant nutrition. It was Liebig's aggressive literary style that turned attention from humus to mineral salts, especially his statement, quoted in Chapter 6, that the crops on a field diminish or increase in exact proportion to the diminution or increase of mineral substances conveyed to it in manure. One can understand that the humus theory, as it then stood, had to be demolished to make way for the mineral theory, but that both were simultaneously tenable could not have been realised by Liebig, or any other worker of this period until a science of soil microbiology had been established. It is now known that, among other functions, humus supplies the elements of carbohydrates and proteins which are thereby synthesised by soil fungi, the fungi constituents then being consumed by the roots in what is known as mycorrhyzal association. Liebig's interpretation of these processes was a purely chemical theory of putrefaction or 'eremacausis' as he named it, by which humus was slowly oxidised to carbon dioxide for use by the plant. It has been said of Liebig that "it did not occur to him that while the humus theory, as then expressed, might be wrong, humus itself might be right".[33]

Liebig's views on the importance of mineral salts received support from several quarters. The Academy of Göttingen in Lower Saxony had offered in 1838 a prize for the best piece of research which would supply an answer to the question whether inorganic constituents in the soil were essential to plant growth. Two pharmacists of Brunswick, A. F. Weigmann and L. Polstorff, won the competition by comparing the growth of plants in sand with those in which the sand had received additions of mineral salts.[34]

Further support came later from Boussingault's experiments on the action of nitrates and phosphates on plant growth in which seeds sown in sterile soil containing mineral salts developed to maturity in spite of the absence of humus or carbonaceous material.

Frédéric Kuhlmann and Putrid Fermentation

Before Boussingault could proceed with such experiments there was an important point which needed clarification, namely, did the plant take up nitrogen compounds from the soil as nitrates or as ammonium compounds? In 1843, Apollomari Bouchardat (1806–1886) claimed that various ammonium salts did not supply nitrogen to the plant but, surprisingly, were even poisonous in dilutions of one part in a thousand.[35] Charles Frédéric Kuhlmann (1803–1881), the very practical professor of chemistry at Lille, working with a field of hay found that ammonium compounds, far from poisoning a crop, actually increased the yield.[36] He used ammonium chloride, sulphate and nitrate; also the ammoniacal liquor from gas-works, horse urine, and, living at Lille, he had available 'night soil' or 'engrais flamand' which contributed much to the success of crop production in Flanders. Kuhlmann also concluded, but without experimental verification, that the roots received ammonium carbonate from the soil and so gained nitrogen. He believed that ammonium carbonate was formed either by the decomposition of nitrogenous fertilisers or by double decomposition between ammonium salts, used as fertilisers and the calcium carbonate of the soil. Kuhlmann's faith in the rôle of ammonium compounds, especially ammonium carbonate, was related to the alkalinity of these materials which was then considered of some importance in a soil. How did Kuhlmann relate his hypothesis to the fact that nitrates were found in the upper layers of the soil? His explanation was that ammonium compounds in the soil were oxidised to nitrates but were then, 'deoxidised' to ammonium compounds by a process of 'putrid fermentation'. Referring to the nitric acid of nitrates in the soil in 1846 he wrote:

> This acid, under the deoxygenising influence of putrid fermentation, undoubtedly passes to the state of ammonia before being assimilated by plants.[37]

He had no firm proof of the presence or action of putrid ferments but referred to various analogous laboratory reactions in which nitric acid was reduced to ammonia by nascent hydrogen in the presence of a platinum catalyst.

Putrid fermentation obviously performed the function of these deoxygenising reactions but there were no laboratory or field experiments to support his hypothesis, the weakness of which was its dependence on the presence in the soil of fermentable organic matter. As was seen much later, Kuhlmann was groping with chemical explanations of phenomena which were being activated by the then unknown soil bacteria.

The supposed rôle of putrid fermentation was tested by Boussingault in 1856[38] when sunflower seeds were sown in pots of sterile soil. Some were watered with distilled water and others with saltpetre solution. The former were poorly developed with a few pale leaves borne on thin stems whilst the others were well developed and each bore a flower. Not only did these experiments demonstrate the striking effect of saltpetre on the growth of a plant but showed that it could develop without the presence of a 'putrid ferment'. For Boussingault there was a more important conclusion which might be drawn from the above experiments, namely, do the roots absorb nitrogen in the form of nitrates, or as ammonium compounds as Kuhlmann believed? Also, is there an interchange between nitrates and ammonium compounds in the soil before nitrogen enters the plant? Firstly, it was necessary to account for the nitrogen which had been supplied by the sunflower seeds and the potassium nitrate with which the seeds had been watered. Did they account for the nitrogen in the plant when developed? Yes, there was an excellent correlation between the initial weight of nitrogen in soil and seeds, and the eventual weight of nitrogen in plant and soil.

Although he had accounted for the nitrogen entering the plant, Boussingault had not elucidated the form in which it entered, either as an ammonium compound or as a nitrate. His practical ingenuity was displayed in devising an experiment in which a plant was grown in a soil containing a known weight of potassium nitrate. His object was to determine the potassium-nitrogen ratio in the soil and in the plant because a conversion from nitrate to ammonium compound in the soil would have reduced the nitrogen content in the plant owing to the loss of ammonia. Ingenious though it was, the state of chemical analysis at the time was not equal to the method adopted and the results were unconvincing although Boussingault offered explanations for the discrepancies. Thus Kuhlmann's assertion that nutrients entered the plant as ammonium compounds remained unchallenged, if incorrect. Kuhlmann is generally known for his 'putrid fermentation' hypothesis linked with the belief in the change in the soil of nitrates to ammonium compounds, and although in error on both points he deserves more than just a passing mention. His memoirs on nitrification of the soil were ten years earlier than those of Boussingault and Ville and displayed a clearer perception of the problems involved than either of them possessed at the time. This claim is borne out by questions which he set out to answer in 1846:

> Is the increase in a crop due to the nitrogen in the fertiliser?
> Is the increase due to the alkaline base of the nitrate or to the nitrogen in the nitrate?
> Are phosphates alone active fertilisers?
> Does the inorganic non-nitrogenous constituent of a fertiliser by itself act as a fertiliser?

These questions were later examined by Boussingault and Ville. Meanwhile, Boussingault had not experienced an untroubled year at Bechelbronn. Reporting to Dumas[39] he mentions the deaths at Niederbronn from cholera which is all around them. He is in the midst of hop-picking from a magnificent crop, but the wheat crop is unsatisfactory – just threshing for seed. Potatoes are abundant but diseased. Oats are doing well, not only for horses for he grew an excellent oatmeal suitable for man, and ends the letter by acknowledging Dumas' long-standing interest in agriculture, hence the domestic details.

Again, he used sunflower in three types of investigation to discover the individual contributions to plant nutrition of calcium, potassium, phosphorus and nitrogen, and were arranged as follows:

(a) No fertiliser whatsoever was used in sterile soil.
(b) Calcium phosphate and potassium nitrate in sterile soil.
(c) Calcium phosphate and potassium bicarbonate in sterile soil.

The combination of phosphate and nitrate in the second experiment produced by far the greatest increase in growth, the necessity of both these materials being first demonstrated by Boussingault in 1857.[40] However, Georges Ville had been conducting experiments with results similar to those of Boussingault in 1856 and 1857 using buckwheat and beans, and for these he claimed priority of discovery over Boussingault. Ville's method differed slightly from that of Boussingault in that he used not sterile soils but artificial soils composed of known mixtures of inorganic compounds normally found in soils.[41] Meanwhile, on November 23rd, 1857 Boussingault announced at the *Académie des Sciences* his conclusions that plants develop poorly in a soil provided with a phosphate only, that they make slightly better progress when a nitrate is substituted for a phosphate, but remarkable development is achieved when the soil contains a mixture of phosphate and nitrate. With some justification Ville claimed priority for these three conclusions and a note from him, read at the following meeting of the *Académie des Sciences* rebutted Boussingault's claim to priority and asked for a committee to examine their rival assertions, a request which had the support of Boussingault. Aware, no doubt, that priority could be ascribed to both of them, Boussingault and Ville agreed that their findings were the result of previous work by scientists and farmers since 1822. There is no record of any such committee being formed; perhaps the *Académie* had recollections of the ineffectiveness of the Chevreul committee on nitrogen assimilation in 1854. It is enough to say that Ville and Boussingault independently discovered the effect of nitrates and phosphates on plant growth, but Boussingault's official claim to priority can be recognised as he was the first to announce publicly his experimental results, a condition required by the *Académie*.

Thereafter, there were no further scientific or personality confrontations as both pursued their own diverse paths. Boussingault worked in such varied fields as the metallurgy of iron and steel, photosynthesis and soil nitrification. Ville continued his investigations of the effects of mineral salts on plant growth and projected them on a large scale at Vincennes, to the east of Paris, where on land granted to him by Napoleon III he became well-known as a pioneer of the use of chemical fertilisers.

Ville was disliked by his professional colleagues at the *Muséum d'Histoire Naturelle*, a political appointment in 1857, yet some of his remarks are those of a droll character, likely to raise a laugh in a lecture theatre of lively students:

> "Lavoisier lost his head but I haven't lost mine." "Certainly my books will be reprinted like those of Molière", and pointing to boxes of unpublished memoirs, "Unless these are published, Science will stand still for twenty five years".

He believed that he was the discoverer of nitrogen and a favourite remark was a paraphrase of one attributed to Louis XIV, "L'azote c'est moi!"[42]

Towards the end of their lives almost all animosity had disappeared, Boussingault making no further reference to Ville while Ville expressed professional admiration for Boussingault though still with reservations concerning his character. "Boussingault was a very intelligent and perspicacious savant of very great ability, but aside from these great qualities to which I have pleasure in paying a tribute . . . he was egotistical and vain surpassing anything one could imagine, and I know that in discussion he was extremely dishonest."[43] Not the character we glean from Boussingault's friends and their correspondence – but these were the remarks of a sick man, for Ville was plagued in his last years by a nervous disease and an abdominal tumour.[44] Both men obtained some degree of affluence and scientific recognition; yet to the end Ville declined to become a member of the *Académic des Sciences*, perhaps because for him imperial patronage was sufficient. Their disputes might suggest that temperamentally they were opposites, yet it is possible that they differed in degree rather than in kind. Whereas Boussingault at times displayed a slight paranoia associated with a schizoid personality, it can be truly said of Ville that he exhibited a strong and continuous paranoia. Perhaps a mutual recognition of each other's weakness prompted some antagonism; a matter best left to psychologists.

Boussingault had established in this short period, 1855–1858, that 'putrid fermentation' and 'putrescible matter' were not necessary in order that plant nutrients could be absorbed by the roots. Kuhlmann, in suggesting this, had worked with a field of grass intended for hay whereas Boussingault eliminated any possibility of nitrogenous organic matter by using the more controllable pot technique. Boussingault had also shown that a successful fertiliser must combine the characteristics of a nitrogenous material with those of a mineral salt. Liebig had erred in insisting that only the mineral content of the fertiliser was of any importance; he did not realise that a preliminary supply of nitrogen was essential if development was to proceed beyond the stage of germination; and also that the rôle of carbon dioxide assimilation by the plant was dependent on the initial assimilation of nitrogen. Boussingault had also made a more critical investigation than his predecessors of the quantitiatve effect of a soluble nitrate on plant growth. His early prescience on the subject of mineral fertilisers was the result of the analysis of soils and stockfeeds which he included in his *Economie rurale* of 1843–44, but not until the appearance of the second edition of 1851 was he able to say:

> I have constantly stressed the point that fermentable organic substances, used to fertilise the soil, should not only be nitrogenous but, independently of the nitrogen, they should contain alkalies, salts and especially phosphates.[45]

The period 1855–58 was one of Boussingault's most fruitful and successful, following as it did a number of apparently unsuccessful attempts to establish the fact of nitrogen assimilation by plants. He had demonstrated the nature of the soil nutrients which entered the plant and how they behaved by themselves, or associated with one another. Atmospheric nitrogen did not enter the plant directly but the soil contained a supply of available nitrogen which could not be accounted for by the amount of fertiliser applied; this he had established by his rotation experiments of 1836–41. There was now the intervening gap in his knowledge of plant nutrition, the nitrification of the soil, a subject which he investigated from 1859 to 1873.

CHAPTER 11

THE NITRIFICATION OF THE SOIL

Up to about the middle of the nineteenth century the soil had generally been regarded as a static system to which fertilisers were added, and which afforded anchorage to plant life, but there were some who believed, without any experimental proof, that the air provided some of the nutriment of the plant. The assimilation of carbon dioxide from the atmosphere had been demonstrated at the beginning of the century and brilliantly publicised by Nicholas Theodore de Saussure (1767–1845) of Geneva in his work. *Recherches chimiques sur la végétation*, with very little effect on agricultural thought as humus was still regarded as the main source of plant carbon. Interest appears to have been transferred to the source of nitrogen, and the atmosphere was a possible answer. In the previous chapter, Jean Antoine Chaptal is quoted as one who thought that plant nitrogen was supplied by the atmosphere but three years previously, in 1826, a M. Longchamp had contributed a paper to *Annales de chimie et de physique* on a new theory of nitrification.[1] He was a former employee of the *Administration des Poudres*, a government department devoted to the improvement in quality and quantity of that essential component of gunpowder, saltpetre, an improvement which contributed greatly to the success of the Revolutionary armies using a purer material and therefore more explosive, an achievement largely due to the efforts of Antoine Lavoisier when he was the Gunpowder Administrator. Longchamp recognised the similarity between a saltpetre bed, or *nitrière*, and a well cultivated field, a similarity noted by Boussingault in 1831 when he visited the *nitrière* of Tacunga in Ecuador. Longchamp overstated his case when he declared that nitrates are formed exclusively from the elements of the atmosphere – partially confirmed later by Liebig's observation that an electric discharge in air would produce nitric acid.[2] But he was correct in stipulating that a successful production of nitrate requires sufficiently humid conditions and a circulation of air. Longchamp's ideas were embedded in a mass of verbiage and his prolix style antagonised his contemporaries who supposed that he was using the debate as an opportunity to attack former colleagues of the *Administration des Poudres.*[3]

Boussingault, in 1857, found that the top soil in his kitchen garden at Liebfrauenberg, in Alsace, contained such an excessive quantity of saltpetre –

over a thousand kilograms per hectare – that he was unable to separate the sugar from the saltpetre in his sugar beet.[4] This was not at all surprising, not only because the garden had been regularly manured for centuries but also because the same materials and care had been bestowed on the garden as was given by saltpetre manufacturers to their *nitrières*. His garden was a *nitrière* without the roof which normally afforded protection from the rain. After a heavy shower he found only a twentieth of the original saltpetre content in the soil but after a further fourteen days of drought this figure had more than recovered itself and was almost one and a half thousand kilos per hectare. If these results were characteristic of cultivated fields he was anxious to compare them with those of uncultivated lands to which end he analysed soils from over a wide area, from the heights of the Vosges to the banks of the Rhine, finding that in spite of rain and lack of manuring they yielded from about one to twelve grams of saltpetre per cubic metre of soil. Expressing the saltpetre content first in kilograms per hectare and then as grams per cubic metre was symptomatic of his indifference to dimensions and even approximations, so that one is mildly surprised to find that for many years he was an examiner for a Statistics prize at the *Académie*. It was obvious that however rapidly saltpetre was washed out of the top soil by rain, it was renewed by a fresh production of saltpetre. Did the renewal of nitrates originate from atmospheric nitrogen, or from the organic matter of manure, or from the water which entered the soil, or yet by the transformation of ammonia washed down by the rain as was believed by Liebig and Kuhlmann?

Boussingault Attempts a Microbial Explanation of Soil Nitrification

Before venturing answers to these questions Boussingault in 1859 summarised existing knowledge,[5] and although it was a valuable synthesis which included priorities where discoveries were concerned, the most important contribution was a recall of Liebig's statements on soil nitrogen:[6]

(a) the amount of nitrogen in the soil was far in excess of that contributed by fertilisers.
(b) in large-scale farming the nitrogen introduced by a fertiliser to the soil was only a fraction of that removed by the crop.[7]

Boussingault's review of current knowledge about soil had compelled him to admit that although the physical state and chemical composition of a soil might be known, knowledge of the soil was very imperfect. As mentioned in Chapter 6, he understood that the vegetative soil (*terre vegêtale*) consisted of (i) decomposed minerals, (ii) organic debris (terreau), and (iii) humus formed by decomposition of the organic debris. He gave a physical definition of *terre végétale* by example:

In running a deepish furrow in a cultivated field we generally distinguish at a glance the depth of the superficial layer which is commonly designated as mould or vegetable earth; this is a layer generally impregnated with humus, and looser and more friable than the sub-soil on which it rests.[8]

It follows that the same substances are found in *terreau* and *terre végétale* but more concentrated in *terreau.*[9] A chemical definition of *terre végétale* was:

. . . the remains of organisms which had lived on earth, ammonium compounds and nitrates, phosphates, alkaline and earthy salts, and nitrogenous organic substances . . . .[10]

A singular omission, of course, was the presence of living things in the soil, but this was soon to be rectified.

During the period of the nitrogen assimilation experiments of 1851–1854, as described in Chapter 10, some plants were enclosed and exposed to filtered ammonia-free air whilst others were grown in the open, protected only from rain and dust. In the latter type of experiment Boussingault had noticed the presence of green growths on the surface of the soil and on the lower parts of the flower pots,[11] for these growths were absent from pots which had been enclosed. They were undoubtedly formed by algae and perhaps were without any effect on nitrification, but they served to draw his attention to an investigation of Armand Bineau (1812–1861), his successor at Lyon, in which fungal moulds, collectively termed *mycoderms*, had appeared on a solution containing ammonium salts and nitrates.[12] The moulds had extracted nitrogen from the salts in solution and Boussingault wondered whether such organisms could fix atmospheric nitrogen. He knew that the soil of a fallow field gradually lost its store of ammonium salts and nitrates and therefore suggested that ammonia had united with humus acids to form stable nitrogenous compounds.[13] He had an alternative biological explanation that the loss of these salts was due to an invasion of *mycoderms*. His 1861 statement was highly prophetic when one considers that *mycoderms* included microorganisms such as fungi and bacteria.

. . . I must ask myself whether the mycoderms themselves are not endowed with the faculty of directly fixing atmospheric nitrogen.[14]

Yet Boussingault was not thinking in terms of continuous bacterial fixation. He suggested that the dead mycoderms contributed their fixed nitrogen to the soil,[15] thus illustrating his conceptual failure to substitute a biological for a chemical solution of the problem of soil nitrification.

At this time, 1857, Louis Pasteur (1822–1895) was investigating the process of lactic fermentation, first undertaken by Scheele in 1780 when he discovered lactic acid by the action of a bacillus, *B. acidi lactici*, on the sugar lactose which occurs in milk.[16] The isolation and identification of this bacillus, which causes the souring of milk, was performed by Pasteur in 1857 and 1859[17] and led Boussingault in 1860 to examine the nitrifying action, if any, of fungal growths in milk whey.[18] At the Conservatoire in May of that year he added acetic acid to milk to separate the fat from the whey which was then divided into three equal portions. To one portion of the whey was added a weighed amount of oxalic acid to neutralise any ammonia present, the solution then evaporated to dryness and the nitrogen content of the residue found, this being, of course, a control experiment. A second portion of the whey was exposed to the air for 42 days during which time there appeared on the surface a *Penicillium* culture. A third portion was exposed to the air for 66 days and both this and the previous portion were treated with oxalic acid, etc., as in the control experiment. Boussingault found, to his great disappointment, that there was a steady decrease in the nitrogen content of the whey as it was left for longer periods exposed to the air. Repetition of the experiments at Liebfrauenberg from August to October 1860 merely confirmed the Conservatoire results. For Boussingault this appears to have been a critical experiment for he was, at that time, prepared to admit of a biological explanation rather than a chemical explanation of soil nitrification if the whey had increased its nitrogen content. The elementary state of microbiology could not then reveal that although fixation of atmospheric nitrogen did occur with *mycoderms*, there was simultaneously a bacterial denitrification of protein in the whey which accounted for the slow but steady loss of nitrogen[19] in the form of ammonia, a loss which was perceptibly noticed by Boussingault during the investigation.[20] It is now known that there are at least six bacilli which react with sour milk to produce ammonia. These discouraging results impelled him to seek a purely chemical explanation of soil nitrification, which was regrettable because he was on the verge of a micro-biological interpretation.

Last Attempts to Provide a Chemical Explanation

One of his last experiments was to enclose soil in bell jars for eleven years (1860–1871), examining the nitrogen, nitrate and carbon content at the beginning and end of this period.[21] Whether such a long interval was intended is not known but during this time he transferred his activities to his son-in-law's factory at Unieux, near Saint-Etienne, and there was also a break in residence at Liebfrauenberg during the Franco-Prussian war of 1870–1871. Three samples of garden soil were used, the first being mixed with sand and water and then covered by a 100 litre bell jar. The second sample was similarly treated but cellulose was added to verify whether the presence of a carbonaceous substance resulted in greater nitrate formation. Liebig's ideas of 1840 had influenced the organisation of these experiments and this particular case recalls Liebig's belief that the slow oxidation of carbon was conducive to nitrate formation in the soil.[22] The incorporation of sand was also in accordance with Liebig's belief that the production of soil nitrates was dependent on a satisfactory air supply. Sand was omitted from the third sample and a smaller, 7 litre, bell jar used to test whether oxygen deficiency inhibited nitrate production. Analysis showed that the total nitrogen had decreased slightly during the eleven years but the result may have been due to denitrifying bacteria. An examination of the air in the bell jars might have shown an increase in nitrogen to balance this loss, for Boussingault was highly skilled in the analysis of air. However, there was an increase in nitrate content in all three cases, but Liebig's ideas were only partially vindicated. The presence of carbonaceous matter in the soil did not accelerate the production of nitrate, although the loss of carbon showed that oxidation to carbon dioxide had taken place. There was a marked decrease in nitrate production in the third and smaller bell jar in which sand was omitted, showing the effect of a restricted air supply. Boussingault's method of estimating the amount of nitrate produced depended on the bleaching action of a nitrate on a blue solution of indigo sulphate, a method due to Henri Descroizilles in 1789[23] but raised by Boussingault to a high degree of precision when, in 1856, he determined the nitrate content of a water sample from the Dead Sea. Otherwise, for an experiment which had lasted eleven years the results may be considered disappointing inasmuch as they did not advance the knowledge of soil nitrification. In a discussion on Boussingault's paper on this subject at the *Société Centrale d'Agriculture* in 1875[24] Pasteur remarked that he was interested in discovering whether infusoria, i.e., protozoa, bacteria and moulds played a part in nitrification but Boussingault, whilst expressing interest in Pasteur's ideas, replied that he was content to examine the conditions under which the top soil nitrified itself, insinuating that he preferred to concentrate on a chemical rather than a biological explanation.

The rôle of atmospheric nitrogen in nitrification had long eluded Boussingault from the time when he first saw the government saltpetre beds of Tacunga in Ecuador in 1831. If the eleven year experiment, just described, was the last chronologically, it was not the last in importance. In 1859 a ten kilogram prism of Liebfrauenberg garden soil was examined for saltpetre content at intervals of 15 days. For six weeks there was an increase in saltpetre but thereafter there was a slow but steady decrease which might have been explained by the action of denitrifying bacteria but for his obsession for chemical explanations. But he had come to the conclusion that the nitrogen of the air did not contribute directly to the nitrification of the soil and suggested that lightning discharges producing nitric acid were the cause.[15] With his predilection for analogy he believed that just as the ocean provided salt, so the ocean of air provided saltpetre; "the aeriel ocean is really a vast nitrière".[26]

When he set up his eleven year experiment he also started a five years investigation (1860–1865)[27] intended to discover whether the chemical nature of the soil was a factor in nitrification. This was his last really serious attempt on the boundary of knowledge. He regarded sand and chalk as important soil constituents but did not include clay, which was a pity because in 1885 Marcellin Berthelot (1827–1907) found that clay soils assimilate nitrogen in the presence of certain bacteria. Boussingault arranged three sets of experiments in which nitrogenous materials were mixed with sand, chalk and garden soil. The nitrogenous materials included such disparate substances as wheat straw, colza oilcake, powdered bone, roasted bone, horn scrapings, wool scraps, dried meat and dried blood of which the nitrogen content of each had been previously determined. The garden soil may have come from Liebfrauenberg but as the sand was from Fontainbleau, the chalk from Meudon and the dried meat and blood from the abattoirs of Aubervilliers, all within easy distance of Paris, it was possible that the experiments were conducted at the Conservatoire. After five years, during which time the soils were watered twice, the sand, chalk and garden soil were analysed for nitrates and ammonium compounds from which the total nitrogen conversion could be ascertained. Only the unburnt powdered bone had any power of nitrification in sand and chalk, this result explaining the high nitrate content of the water which flowed from the Paris Catacombs, the resting place of over a million skeletons which, in 1785, were transferred from various cemeteries, notably that of the Church of the Holy Innocents. Although the greatest conversion of nitrogenous material into nitrates and ammonium compounds occurred with garden soil a possible biological explanation was ignored and Boussingault could proceed no further with the problem of nitrification.

From 1876 onwards he produced a number of memoirs on South American topics; he returned to the investigation of barium peroxide which he had left in 1852, studied the functions of leaves and was interested in the metallurgy of chromium and nickel – indeed a polymath to the end.

He had constructed a rough, though workable, model of the nitrogen cycle. Organic material in the presence of the top soil, or *terre végètale*, yielded ammonium compounds and nitrates, the ammonium compounds being oxidised eventually to nitrates which are plant nutrients. Nitrogen in manure, the dead remains of plants and animals, furnished nitrogenous material to be converted again by the *terre végètale*. Yet the intermediate steps were missing, and not only was the new science of microbiology required but also a new breed of chemists with fresh conceptions of chemical processes in nature based on the fermentation discoveries of Pasteur. At 74 years Boussingault could hardly be expected to reorientate his well-established chemical tradition.

A Bacterial Explanation Appears

Meanwhile, microbial investigations were slowly making progress. In Britain clay pipes had been used since the middle of the century for draining lands to bring them into a fit state for cultivation. In 1856 J. Thomas Way (1821–1884) examined the ammonia and nitrate content of land drainage water from farms in England because it had been suggested that the benefits of drainage might be offset by nitrate loss in drainage water.[28] Way was surprised to find that in such water, the amount of ammonia had decreased whilst that of nitrate had increased, but he wrongly ascribed the result to the heavy manuring of the fields for he was unwilling to believe that ammonia had been converted to nitrates in the soil, and even suggested an enquiry into the possible production of nitric acid from manure.[29]

The first intimation that micro-organisms were responsible for soil nitrification came from a German scientist, A. Muller, who in 1873 commented on the stability of sterile ammonium salt solutions in contrast to the disappearance of ammonium salts in natural waters.[30]

In 1877 Theophile Schloesing (1824–1919) and Achille Müntz (1846–1917) wondered whether nitrification was the result of a purely chemical and direct interaction between oxygen and nitrogenous compounds, or whether it was the result of intermediate organisms functioning as ferments.[31] They were engaged on studies on the purification of sewage water and allowed this material to seep through a metre-long tube, packed with an incinerated mixture of sand and limestone, in the presence of oxygen. After eight days, when the first of the sewage water filtered through, the filtrate contained ammonia, and continued thus for twenty days. Thenceforth, nitrates appeared and increased in quantity. The fact that it took such a long time for nitrates to appear convinced Schloesing and Müntz that the process was not one of simple oxidation which would have occurred at once. After nitrification had proceeded for four months, chloroform was passed through the tube; ten days later the filtrate contained ammonia but no nitrates and nitrification was restored only by the introduction of soil washings. Schloesing and Müntz concluded that the conversion from ammonia to a nitrate was a biological process due to "le ferment nitrique" which was then the term for the nitrifying bacteria. These results confounded Liebig's theory that nitrification was a purely chemical oxidation of ammonia.

R. Warington[32] and J. Soyka[33] in 1878 were the first to demonstrate that nitrification occurs in at least two stages – the formation of a nitrite from ammonia, and the conversion of the nitrite to a nitrate – but neither could isolate the organisms responsible for these changes. The then recently introduced technique, due to Robert Koch (1843–1910), of culturing bacteria on gelatine plates proved unsuccessful because, as was later discovered, the nitrifying bacteria, although devoid of chlorophyll, flourish in an inorganic medium. The nitrifying bacteria of the soil, *Nitrosomonas* and *Nitrobacter*, were isolated in 1890 by Winogradsky using an inorganic medium consisting of a solution of ammonium sulphate and potassium phosphate.[34]

Thus, after Boussingault's many and skilful attempts to demonstrate nitrogen fixation and to explain the nitrification of the soil, the final answers were supplied by others. Boussingault was ageing a little physically, but certainly not mentally, yet there were several serious reasons for his inability to complete this work. He was unable to appreciate sufficiently the work of his younger contemporaries, and although he was filled with admiration for the laboratory techniques of Pasteur he essayed to explain his work in terms of the scientific field with which he was familiar, and was thus unable to appreciate the new concept of reactions initiated and controlled by microorganisms. Endeavouring to explain cryptogamic growth in terms of phanerogamic conditions, i.e., the conditions of ordinary plant life, he would insist that micro-organic growth was possible because rain-water contained ammonium salts, nitrates and phosphates,[35] as if cryptogamic growth was merely a question of nutrition and not a complete chemical change of the nutrient medium. The contrast between the two viewpoints is illustrated by Boussingault's reminder that before sugar could be fermented to alcohol and acetic acid it was necessary for the sugar to be produced with the aid of chlorophll and light. True enough, but Pasteur maintained that without micro-organisms the work of death could not be completed, and that without them the earth's surface would be littered and encumbered with dead organic material making life impossible.[36] Although his son Joseph was at that time engaged in analysing a vast number of wines for the *Exposition Universelle* of 1878 it appears to have evoked no interest for Boussingault who continued to have no definite explanation of fermentation and for this reason he was not as didactic as Liebig, who regarded, erroneously, fermentation and decay as merely processes of general combustion with the oxygen of the atmosphere.[37]

CHAPTER 12

EDUCATION AND HYGIENE

Although today the interrelationships between subjects is acknowledged, this broad view was not always recognised and the discussions at the *Conseil de Perfectionnement* of the Conservatoire illustrate this tendency of maintaining strict boundaries between even related subjects. Towards the end of each year it was customary for the staff to submit their syllabuses for the following year to be read and commented on at the Council meetings. Boussingault usually managed to escape from any criticism of his syllabuses by postponing his return from Alsace until November or as late as December. In 1853, at an October meeting of the Council,[1] his syllabuses were strongly criticised by Baudement, professor of applied zoology or *zootechnie* as it was called, an unfortunate term coined by Adrian Gasparin, the first Director of the *Institut agronomique* at Versailles, who regarded a farm animal as a machine burning fuel and giving in return meat, milk, etc. Not so, Baudement, whose courses in pure zoology were drawing crowds of students to his lectures which were of a traditional nature and certainly not technological. Baudement objected to Boussingault's treatment of 'animal tissues and animal secretions', suggesting that his teaching should be confined to the chemical aspects of these topics, and also that 'the production of butter and cheese' was the concern of the applied zoology course. The president of the Council, at that time Peligot, left the revised drafting of the course to Boussingault's discretion but when, at the next meeting,[2] a more detailed syllabus was read, Baudement again objected with the naive remark that "the syllabus includes all that can be said on the feeding of cattle from the chemical and physiological point of view", indeed a compliment to Boussingault who attended the following meeting[3] declaring that he had no intention of encroaching on the syllabuses of anatomy and physiology but that he could not be confined to boundaries that were difficult to define, and if defined would impede the liberty of the teacher. That which was left unsaid, and tactfully so, was that Boussingault's chair had been changed in the previous year from that of agriculture to agricultural chemistry as a result of the foiled attempt to oust him in favour of Ville; yet Boussingault continued to teach agriculture.[4]

A different subject was the target of criticism in 1855[5] when the Minister of Commerce and Industry met the professor of industrial economics, Burat, and not only suggested that certain portions of his syllabus were dangerous but informed him that he, the Minister, had decided to change the title of the chair to Administration and Industrial Statistics. Burat was worried because the new syllabus omitted such essential though controversial topics as Work, Money, Banks, Organisation of Credits, and Customs. "How can such a course exist if it does not treat of these subjects?" protested Burat. A disappointing and unrealistic contribution from Boussingault conceded that industrial economics was a science but that it was sufficiently covered by the syllabuses of his colleagues. Outside the subject of *économie rurale* it appears that Boussingault was not well-informed on economics in general for again, at a Council meeting on 26th April, 1864 Boussingault, among others, expressed concern at the teaching of political economy, but Wolowski who held this chair thought that his subject helped to combat "the poison which, in the form of Communism, seeps into the workshops and the press".

That lack of consultation among staff was characteristic of the Conservatoire is demonstrated by several instances which were essentially professional rather than political. One would have expected Arthur Morin, the Administrator of the Conservatoire, to have had an impartial and objective attitude towards syllabuses but he questioned the physics professor's treatment of 'pumps and the hydraulic press'.[6] He held that the professor should confine himself to the phenomena and refrain from discussing the mechanical effects such as the applications of these machines, claiming that the applications of pumps was the preserve of the professor of applied mechanics. Again, in October 1859[7] we find Morin complaining that the subject 'water' was being dealt with by three professors, Pelingot, Boussingault and Persoz who was now on the Conservatoire staff. Morin maintained that Peligot, alone, was entitled to treat this topic in a general manner and that the others should explain its properties and reactions with reference to their particular courses, which in the case of Persoz referred to textiles and dyeing. And so Boussingault would postpone any teaching on the mulching of plants until Peligot had explained the phenomenon of capillarity!

That collaboration could occur at times of collegiate crisis was shown when Baudement died in 1864 leaving vacant the chair of Applied Zoology.[8] The Minister of Commerce and Industry asked the Council to consider whether the chair should be continued. Boussingault, the first to reply, took the usual Conservatoire line that all the necessary zoology was being taught in the department of agriculture, and that therefore the chair should be abolished. As Baudement had been a popular lecturer there was sympathetic support for the retention of the chair but eventually the latent conservatism of the Council won the day, but not without some bitter and facetious criticism from Emile Trélat, the professor of civil engineering, who disagreed with Boussingault because a course in applied zoology must perforce include the theoretical background so essential to an understanding of the practical applications. Morin wisely delegated the problem to a sub-committee consisting of the two professors of agriculture – Boussingault and Moll – with Morin and Henry Tresca – professor of mechanics – which rapidly produced a report in time for the following week. But the sub-committee was prejudiced, damning applied zoology with faint praise, but suggesting that it could best be taught by a union of the two agricultural courses together with a third course on 'agricultural operations and farm engineering'. The recommendations were part of a *Report on the Organisation of Agricultural Education at the Conservatoire* which when debated at a Council meeting[9] evoked facetious criticism from Trélat, as the Conservatoire had neither the means nor the opportunity to practise agricultural operations and farm engineering. By 15 votes to 3 the sub-committee's proposals were carried, Boussingault and Moll voting with the majority asking for the abolition of the chair. Clearly, the Conservatoire professors had developed a parochial stance towards their subjects, partly because they were on the defensive against syllabus revision, but this was not the principal reason. The rise of new subjects in the realms of technology, sociology and economics presented a grave challenge to those who held chairs in the well-established disciplines, for demands were made for the establishment of additional chairs which could only be implemented at the expense of existing chairs.

Institut National Agronomique

The history of the development of higher agricultural education in France is inseparable from the achievements and vicissitudes of the *Institut National Agronomique* established at Versailles in 1850[10] resulting from a debate, mentioned in Chapter 9, in the *Assemblée Nationale* in 1848. Sited on three royal farms of 1500 hectares, instruction was given in the outbuildings of the famous château.[11] It attracted a well-qualified staff which included the chemist Charles Wurtz (1817–1884), the botanist Pierre Duchartre (1811–1894) and the Director was Adrian Gasparin who was Préfet of the Rhône département when Boussingault was in Lyon in 1835. The Institut fell with the Second Republic in 1852,[12] the official reason being the cost of maintenance but the true explanation was that the farm interferred with Napoleon III's hunting when at Versailles.[13] Gasparin, with his flair for summing up a situation in lucid terms, said:

It is by the head that we educate Society; the lamp must be placed on high so that it can be seen from afar; a school like the Institut Agronomique, in sending out skilled men throughout the country, could have done more in twenty years for the progress of agriculture than ten generations of little schools where one worker in five thousand is taught incompletely.[14]

Boussingault was a member of the agricultural commission, *l'Enquête Agricole de 1866–1870*, established as a result of an Imperial decree of March 1866 of which J. B. Dumas was chairman and Eugène Tisserand (1830–1925) was secretary. Tisserand was a former student of the *Institut Agronomique* and whilst Inspector General at the Ministry of Agriculture he campaigned for its re-establishment. On a sub-committee formed to establish a successor to the defunct *Institut* sat Boussingault. Terms of reference also included ways and means of encouraging students to make agriculture their career[15] but any suggestions of a pragmatic nature were doomed whilst Boussingault insisted on locating the course at the Conservatoire and whilst others suggested the *Ecole Centrale* and the *Ecole Polytechnique*, all three well inside Paris without any farm-land available such as existed at Versailles. Yet credit must be given to Boussingault who candidly admitted the shortcomings of the courses at the Conservatoire:

It is true that there are already at the *Conservatoire Nationale des Arts et Métiers* courses in agricultural science, agriculture and rural economy, agricultural technology and lastly, a course in agricultural chemistry, taught by me, but it is difficult to say to whom the different courses are directed, and in any case the agricultural profession is hardly represented.[16]

He further maintained that students should be recruited from among the sons of major farmers, for "agriculture does not offer sufficient openings to those who are not in this privileged position".[17] A methodological flaw accounted for his reactionary stand as stated in his *Economie rurale* of 1843 that the theory and practice of agriculture could not be taught simultaneously.[18]

It was a great pity that the Conservatoire 'lobby' got its own way with the Minister of Commerce and Industry, who fell for a 'hot seat' policy in which the *Institut Agronomique* students used the lecture theatres and laboratories when not used by the Conservatoire. The technical library and laboratories were, of course, an inducement to restart the *Institut* at the Conservatoire, but in addition there was the collection of models, dating from the originals of Vaucanson, and mentioned intermittently in the minutes of the Council meetings, giving the impression that they were almost sacred objects, or at least they corresponded to the 'ideas' or 'forms' of Plato, perfect examples compared with which the everday copies were inferior. But the Conservatoire 'lobby' was also aided by the need to reconstruct the French army after the disasters of 1870–1871 for which finance could only be obtained by ruthless economies.

The Conservatoire and Industry

Towards the end of a Council meeting, at the beginning of 1860, Morin announced that he was preparing to take certain tentative steps towards a link with industry because some chemical manufacturers wished to establish a laboratory at the Conservatoire for the testing of their products.[19] It might have been assumed that the staff would have welcomed this gesture on the part of industry but Boussingault and his colleague Peligot suspected that the project, innocent though it seemed, was fraught with difficulties. Postponed to the next meeting the discussion was led by Boussingault who expressed himself at great length against the proposal, his main contention being that the creation of a special laboratory under the control of Conservatoire staff would involve them in a moral obligation to conduct or encourage the lines of research demanded by industry.[20] Further, the Conservatoire was well provided with chemistry laboratories and one extra would cause the subject to loom too large in the college programme. Although not his principal objection, Boussingault reminded the Council that the Conservatoire already provided an advisory service to industry but Morin's proposal could mean an attempt to install a chemist who was a nominee of industry. A minor point which he also raised was that demonstrators would lose their analytical fees paid by industry, although these amounted to little. Morin, with his more balanced and objective approach, appealed to all professors to find the means of putting this project into practice. Altogether, the staff appear to have given the impression that technical education was the preserve of an élite such as themselves and characteristic of an in-looking group which was already ageing.

Conseil d'Hygiène

A biographical enigma is the frequent reference to Boussingault's work for a watch-dog committee on health and hygiene, the *Conseil d'Hygiène Publique et du Salubrité*, such as the statement by Louis Troost at a funeral oration for Boussingault in May 1887, in which he mentioned the various health and hygiene topics which received his attention throughout almost half a century.[21] A *Conseil du Salubrité* was established by Napoleon in each département in 1802, and in 1848 it was called the *Conseil d'Hygiène Publique et du Salubrité* but for brevity it will be referred to as the *Conseil d'Hygiène*. It dealt with the maintenance of public health in each département, the prevention of disease, the prevention and elimination of odoriferous industrial waste, and the protection of workers in industry, especially those exposed to the toxic effects of phosphorus, mercury, arsenic and lead. Each *Conseil d'Hygiène* had executive power supported by various judicial bodies. One of the few documents of the *Conseil d'Etat* which survived the burning of its library in 1871 by the Communards of Paris concerned such executive powers and is dated 25th July, 1851.[22] A Monsieur Miguet owned a tallow factory at Fontainebleau where animal fat was melted over an open fire. It may have been the fire hazard or perhaps the offensive smell, but there had been a complaint to the *Conseil d'Hygiène* as well as to the Commissioner of Police and the mayor. The *Conseil d'Etat* gave Monsieur Miguet six months in which to put an end to the nuisance after he had defied the *Conseil d'Hygiène*, the police and the mayor, but the sequel will never be known because the *Conseil d'Etat* was dissolved in the following December as a result of Louis Napoleon's coup d'état.

That Boussingault took an early interest in the health and well-being of people is evident in his writings as in the papers which he contributed on goitre in the high Andes, the salt mines which yielded traces of iodine, the state of the air and especially the presence of 'miasmas' and their supposed effect on the health of communities. It is not surprising that his association with the *Conseil d'Hygiène* of the Seine extended over a period of 47 years including a vice-presidency in 1855. The *Conseil d'Hygiène* of the Seine issued a detailed annual report which was generally published from a year to three years late, the worst examples being the reports of 1849–58 which were published in one volume in 1861, but it is only fair to mention that the 1845 report was issued in 1846 though this was a notable exception. As the multitudinous reports had of necessity to be brief, it was impossible to assign responsibility for any particular contribution. Yet they are extremely interesting for they represent an aspect of nineteenth-century France not found in the life of château or parliament, but are a register of the growing industrial life of that country with its accompanying dangers, and of the attempts to limit the spread of disease. At a workshop in the rue Saint-Martin, urine was used in the dyeing of animal skins. During the hot summer months this process gave rise to foul odours in the neighbourhood which included the Conservatoire des Arts et Métiers. The *Conseil* asked for ventilation to be installed to carry away the ammoniacal fumes, and for the waste urine to be conducted away by an iron pipe into the drain which went along the rue Saint-Martin.[23]

One of the momentous problems for that year, 1840, was to advise on the exhumation and transport of the remains of Napoleon from St Helena to the Invalides in Paris. There was much erudite discussion as to the possible state of the corpse, for Napoleon had been dead for almost twenty years, but the body was found to be in a good state of preservation.

Explosions and fires were also the concern of the *Conseil d'Hygiène*, many of them due to matches which were liable to detonate spontaneously as the heads consisted of a mixture of chlorate of potash and red phosphorus. Gasometers occasionally exploded. A more modern note, on sound pollution, was expressed by an item in 1845 in which excessive noise was regarded as detrimental to health. Steam engines were often a nuisance to the neighbourhood because of the noise they made. Complaints were often made when the engines were sited in residential buildings. In one case of this kind the *Conseil* asked that the engine be separate from the party walls.[24]

One could continue almost *ad infinitum* to select gems of interest from these reports but before turning to Boussingault's personal contributions, there is just one further example entitled *Danger des Montgolfières* of 1851. *Montgolfières* were hot air balloons similar to the one constructed by the Montgolfier brothers for the first manned balloon ascent in 1783. The attention of the police had been called to the dangers attending balloon ascents at fêtes when air was heated by stoves or by burning tow soaked in alcohol. The resulting fires had burnt forests, thatched houses, crops and stocks of wood. The powers of the police at the time were limited to an examination of the envelope for air-worthiness, especially the stitching!

Boussingault's Reports

The number of reports submitted by Boussingault to the *Conseil d'Hygiène* is remarkably small for such a long period. Undoubtedly his wide scientific interests were of inestimable value in discussions, concerned as they were with an interdisciplinary subject. Yet absence for half the year in Alsace limited his value to any council or committee. The biographical enigma referred to above, arises from the fact that none of his researches concerned the subject of hygiene, with the exception of the nitrate determination in the water of the Seine and in Paris wells. In the funeral oration delivered by Louis Troost reference is made to his work in the conversion of town waste, cess-pool water and 'night soil' into fertiliser, and although this was an absorbing topic with Boussingault, there is no evidence that his researches contributed anything to these subjects.

In 1864 he was called upon to report on a plan submitted to the municipality of Paris by a Monsieur D. who was responsible for the plantations of the city and had proposed that they should be watered with a solution of the well-known hygroscopic substance, calcium chloride. On the basis of a little knowledge Monsieur D. maintained that one watering with the solution would avoid the need for ten subsequent waterings because calcium chloride would absorb the necessary water from the atmosphere. Boussingault had to explain that the object of watering a plantation is not to keep the soil moist but to supply water to the plants. The result would be a plantation in which plants withered in a perpetually moist soil.[25]

In 1866 he examined a so-called 'fertiliser' which had as its object the destruction of cut-worms, or cockchafer larvae, that bite off the shoots of young seedlings. For his raw material, the inventor used the waste from yards where hides were prepared for tanning, mixed with sulphuric acid, alum, soda and common salt. Boussingault found that the insecticidal property of the 'fertiliser' was due to orpiment, a sulphide of arsenic, which was used in yards where skins were prepared. "It is not the first time", commented Boussingault laconically, "that a poison has been used successfully to destroy harmful insects".[26] Another report was on the subject of complaints from the residents in the neighbourhood of a factory at Pantin, to the north-east of Paris, producing a fertiliser, 'taffo', a mixture of town and tannery waste, faecal waste from cess-pools, formed into briquettes and sprinkled with slaked lime or gypsum. After a visit to the factory, where he was most impressed by the attempts to return waste nitrogen to the soil, the complaints were dismissed, but perhaps his zeal for conservation swayed his judgement, for a year later another *Conseil d'Hygiène* member concluded that the complaints were justified.[27] These were individual contributions but with other members various investigations were made, for example at a factory at Aubervilliers, to the north of Paris, making ammonium sulphate from the contents of cess-pools; another converting blood into fertiliser; the sale of koumiss, an intoxicating drink made from fermented mare's milk; tests of beers and ciders, and the state of cheese warehouses.

The Liqueur Kirsch Provides Some Problems

As an Alsation by adoption, Boussingault was particularly interested in the preparation of the the liqueur kirsch, or kirschwasser, obtained by the fermentation of black cherries and subsequent distillation. Some of the cherry stones are crushed with the fruit with the liberation of the glucoside amygdalin which is broken down by an enzyme, emulsin, present in the kernels, to oil of bitter almonds and prussic acid, both toxic in large quantities, but in kirsch the very small amounts impart the pleasant and characteristic flavour of this liqueur. About 1864 it was discovered that a fraudulent kirsch was being produced in the hinterland of the French Riviera, in the Alpes-Maritimes. Cherry-laurel leaves were being distilled, the aqueous solution then mixed with alcohol and sold as kirsch[28] – not only fraudulent but dangerous, as samples were found to contain four times the normal amount of prussic acid. As Boussingault was to find later, there was no easy method of distinguishing between the fraudulent and the genuine.[29] The commercial test was to add tincture of guaiacum, a resin obtained from a West Indian tree, lignum-vitae, when a true kirsch was believed to turn blue. Boussingault found that this was not a reliable test but merely indicated the presence of copper acetate or copper cyanide in the liqueur. Their presence depended on the mode of distillation in the copper stills, country distillers using an open fire, but if distilled carefully on a water bath there was no formation of copper salts. A distiller at Luxeuil in the Haute-Saône complained that his first-class kirsch, although carefully distilled, was shunned by the buyers, but Boussingault's research supported the distiller's contention that, with the buyers, the guaiacum test was more an article of faith.

The subject of kirsch became an obsession with Boussingault in 1874 for in his analysis of samples he had concluded that they contained a dangerous proportion of copper, and he reported to Dumas[30] the need to inform the Minister concerned or the Préfet of Police. Boussingault was in favour of stopping the sale of kirsch. Dumas was not impressed by the alleged gravity of the situation nor was he alarmed as the amount of the copper found did not appear dangerous. Too much family discussion may have caused the copper scare, for the subject had been raised by son Joseph when making a detailed study of the fermentation of stone fruits.[31] On a lighter note,

Boussingault admitted surprise on finding no trace of copper in his agricultural students!

Further Boussingault Reports

As well as the scientific expertise which he displayed, he was also outstanding as secretary to the *Conseil d'Hygiène* in 1843 when he opened the Annual Report with a survey of its functions. The usual three to six pages of schedule of matters before the Conseil was reduced to a single page by making use of fewer categories of complaints and by eliminating the doubtful difference between 'justified' and 'unjustified' complaints. A special note was made of industries mentioned for the first time. The improvement on previous reports is not only noticeable but striking and justifies Troost's tribute, previously mentioned, to a remarkable clarity combined with profound knowledge which was a sure guide to all questions concerning the hygiene of workers.

To the south of Paris, near the parc de Montsouris, is a large reservoir constructed during the years 1868–78 by Eugène Belgrand to receive water from the river Vanne 136 kilometres distant. The reservoir and its aqueduct were part of an ambitious plan to supply Paris with potable water, for during his exile in London, Napoleon III had been impressed with the squares, parks and distribution of clean water to all the inhabitants. A complementary part of the Belgrand plan was the construction of four huge underground galleries into which flowed the waste water of Paris which was directed to the outskirts of the city. Prior to this, the sewage and waste water were conducted by 155 kilometres of malodorous sewers and emptied into the Seine. The impetus behind these colossal schemes was supplied by Boussingault's analyses of water samples from the Seine and from wells, the water of which was used by bakeries. If the Seine was polluted, even more so were the wells which could not remain clean in the presence of about sixty thousand cess-pools. Boussingault found that the nitrate content of bread was high as the result of decomposition of organic matter in the water. He was not unduly perturbed except that he found harmful organisms in bread which had survived the temperature of baking. "One cannot understand why in Paris we continue to make bread with dirty water",[32] he wrote.

His advice was sought on the central heating of the new Paris gaol, the Nouvelle Force;[33] it was a question of deciding between hot water or steam, and steam won the day. This was not the only hygiene problem at the gaol;

there was the odour from twelve hundred sanitary buckets. In 1845, a Monsieur Siret had proposed the use of a disinfectant powder to be placed in the buckets each day, composed of charcoal and metallic sulphates which trapped the offensive ammonia and hydrogen sulphide gases. Boussingault willingly recommended the method, so reminiscent of that used by an Alsace neighbour for converting cess-pool water into fertiliser.[34]

His opinion was sought on two interesting legal cases[35] concerning the manufacture of dyestuffs. The purple natural dye known as orchil, from a lichen *orchella*, was obtained by a patented method in which the lichen was treated with fermented urine, the ammonia of which was the reactive agent. A costly phase of the process was the transportation of the lichen to a point where urine was available in quantity, usually a town. A chemist, Pierre Robiquet in 1829 had discovered a method of extracting orchil from the lichen *in situ* by treating it with ammonium hydroxide, and a commercial process on these lines had been established until a court injunction prevented any further manufacture. A committee which included Boussingault successfully appealed against the injunction.

The second case was that of a Paris firm manufacturing dyes from logwood, the heartwood of the leguminous tree *Haematoxylon campechianum* found in Mexico and South America. A gasworks later established nearby caused tarry matters to pollute the wells of the dyeworks, resulting in a deterioration in the standard of the dyes. The task of Boussingault and his colleagues was rendered more difficult by the gas company's unscrupulous representative, a chemist, Monsieur Salvetat who alleged that the presence of aluminium and iron salts in the wells, which acted as mordants, was responsible. Boussingault's analysis showed not a trace of aluminium and concluded that the inferior dyes were the result of tar in the wells.

This chapter of miscellaneous but related items serves to illustrate various tendencies and characteristics of Boussingault. Firstly, his attitude towards agricultural education was conservative rather than missionary, believing that the Conservatoire must be the only centre for higher agricultural education, and this stand was related to his belief that the theory and practice of agriculture could not proceed together at the same institution. Indeed, had he agreed that these two aspects were mutually compatible, there would have been some justification for the establishment of an agricultural institution in the country, in direct competition with the Conservatoire. Next, in spite of an earlier love of liberty, as exemplified in South America and also in the 1848 elections to the Assemblée Nationale, he was somewhat suspicious of the growing impact of sociological and economic subjects in the curriculum of the Conservatoire. Perhaps he subconsciously realised that the content of these new subjects, and the questions they raised in the minds of people, presented a challenge to his feudal way of life in Alsace, although he merely accepted this feudalism without wishing to take advantage of it.

Lastly, his outstanding secretaryship of the *Conseil d'Hygiène* in 1845 gives cause for regret that such ability was rarely used, for one could have imagined Boussingault as an extremely capable *Sécrétaire perpétuel* at the Académie des Sciences, had he been appointed. His polymathic interests made him an excellent member of the *Conseil d'Hygiène* which required of its members not only a broad horizon of knowledge but the initiative to render the *Conseil* an executive as well as an advisory body.

CHAPTER 13

PHOTOSYNTHESIS, OR THE PLANT AND CARBON DIOXIDE

Photosynthesis is the process by which the inorganic substances carbon dioxide and water are converted to organic compounds in the plant with the aid of light energy, liberating at the same time the gas oxygen, and therefore is the most important process in the world for by its means the plant synthesises food for man and animals, including those on his farm and in his fishing grounds, and those which he hunts. In nature it proceeds with apparent ease but under laboratory conditions the process is not easy to reproduce. Atmospheric carbon dioxide, of which there is about 0.03 per cent, enters the minute holes, or stomata, of the leaf which were first observed by Marcello Malpighi (1628–1694)[1] a pioneer in the application of the microscope to the study of animal and plant structure, and who suggested that leaves were concerned with plant nutrition.

To appreciate the advances made by Boussingault in the realm of photosynthesis, its early development must first be traced. Edme Mariotte (c. 1620–1684) first mentioned the effect of light and referred to "the brightness of sunshine necessary for the feeding of plants".[2] The English minister, Stephen Hales (1677–1761), believed that "Plants very probably draw through their leaves some part of their nourishment from the air".[3] It was Joseph Priestley (1733–1804) who found that a sprig of mint, placed in a jar of air in which a candle had burned, restored the air in ten days so that another candle burned perfectly well in it.[4] This was a very important discovery, a momentous one indeed, but the implications were not realised by Priestley who attempted to explain the phenomenon in terms of the phlogiston theory. That the oxygen set free comes from the water in the plant was first suggested by Claude Bertbollet in 1785 "because light combines with the dephlogisticated air (i.e., oxygen) and imparts to it its gaseous form".[5] From these statements it can be appreciated that an explanation of the basic reactions in photosynthesis was not possible until the oxygen theory of combustion, due to Lavoisier, had supplanted the phlogiston theory.

The Dutch physician Jan Ingen-Housz (1730–1799), whilst living in England, discovered that photosynthesis takes place only in light. He also showed that plants are giving off carbon dioxide continuously.[6] The librarian of Geneva, Jean Senebier (1742–1809) was another who demonstrated that oxygen was exhaled by the plant although in 1782 he referred to it as dephlogisticated air.[7] The weaving together of the various threads to form a true pattern of photosynthesis was carried out by Nicholas Theodore de Saussure in the first modern work on plant nutrition, *Recherches chimiques sur la végétation*, published in Paris in 1804.[8] De Saussure had the advantage of interpreting the changes taking place in photosynthesis in terms of the oxygen theory of combustion. He was the first to compare the volume of carbon dioxide absorbed by a plant with the volume of oxygen set free, and also found that a plant gains in weight when exposed to light and carbon dioxide due to the simultaneous taking up of water in a form not removable by drying. A stage had been reached when photosynthesis could be represented thus:

$$\text{Carbon dioxide} + \text{Water} + \text{Light} \rightarrow \text{Organic matter} + \text{Oxygen.}$$

After this bright start there was almost a hiatus in the chemical and physiological investigation of photosynthesis which lasted until 1861 when Boussingault published his first paper on the subject. Meanwhile two chemists, Pierre Joseph Pelletier and Joseph Bienaimé Cavantou worked together on the green matter of plants which in 1818 they called chlorophyll.[9] Visitors to Paris will perhaps recollect the statue of a reclining sick woman in the boulevard Saint-Michel, a memorial to the discovery of quinine by them. There was a neglect of the findings of de Saussure, Ingen-Housz and Senebier for several reasons. There appeared to be adequate satisfaction in explaining the nutrition of plants by reference to the products of humus which, existing in large quantities in the soil, seemed a more likely nutrient than the feeble quantity of carbon dioxide in the atmosphere. Also, the concept of 'vital force' was still prevalent, indeed there was a resurgence of 'vitalism' at the end of the eighteenth century due to the writings of Marie François Bichat (1771–1802) who postulated that in living things there is a conflict between vital forces and those of physical science by which the forces of physics and chemistry become uppermost after death and the dead remains are destroyed. Unfortunately, vitalism tended to make intellectual effort and empirical investigation superfluous. It was also a period of confusion due to the accumulation of data about the minutiae of form and structure whilst function was neglected, as there was a failure to recognise the importance of the living processes of plant life. Differences in structure were not recognised as due to the possible differences in function because vegetable physiology as a science had hardly come into existence. An exception could be made of of the botanist Henri Dutrochet (1776–1846) who showed in 1837 that in a plant there was an ascending and a descending sap, and he refuted the notion of a circulation of 'vital sap'. He also showed that respiration proceeds by both absorption of oxygen and the decomposition of carbon dioxide taking place simultaneously in light, and therefore these are not nocturnal and diurnal functions as was previously supposed. He established that only in cells containing chlorophyll was carbon dioxide absorbed and oxygen evolved.[10] Mention should also be made of Hugo von Mohl (1805–1872) of Tübingen who in 1856 showed that the mechanism of the stomata depended on the rigidity of the guard cells, the two cells which open and close each stoma. He also observed starch grains in the chloroplasts, the small rounded bodies in plant cells which contain chlorophyll, although he did not realise that the appearance of starch grains was a sign of photosynthesis.[11]

In the realm of the physics of photosynthesis there was a fundamental advance in 1845 when Julius Meyer (1814–1878), a German surgeon and physicist, applied his law of the Conservation of Energy to vegetative processes. In 1842 he had announced the thermodynamic principle that the various forms of energy were mutually convertible and three years later he concluded that Nature had set for itself the task of seizing sunlight in its flight as it streams upon the earth, and thus accumulating the most swiftly moving of all forms of energy by transforming it into a potential state. To this end the surface of the earth had been covered with living organisms that absorb sunlight into themselves and thus generate a permanent store of potential chemical energy. These organisms making up the plant world form a reservoir in which the fleeting rays of light are caught and cleverly stored for future use.[12] Meyer's contribution can be briefly stated by comparing the previous equation, representing the state of knowledge before 1845, with the following:

$$\underset{\text{dioxide}}{\text{Carbon}} + \text{Water} + \text{Light} \longrightarrow \underset{\text{matter}}{\text{Organic}} + \text{Oxygen} + \underset{\text{energy.}}{\text{Chemical}}$$

The chemical energy is stored in plant foods which are eaten and transformed into mechanical heat and energy in the animal. This last equation is a brief but fundamental representation of the process of photosynthesis, and that which has been discovered since 1845 has been an elaboration of the equation, either as a whole or in part. Research, either quantitative or qualitative has served to elucidate various aspects of the equation among which are the investigations of Boussingault from 1861 onwards.

Boussingault and the Photosynthetic Quotient

Boussingault's outstanding achievement in the study of photosynthesis was his series of investigations into the ratio between the volume of carbon dioxide assimilated and the volume of oxygen evolved by a plant in the presence of sunlight, a ratio known as the photosynthetic quotient in which the numerator is the volume of oxygen. His interest in photosynthesis dated from 1854, the period of the nitrogen assimilation experiments, when in a letter of June 23rd to Dumas he mentioned the emission of oxygen from plants immersed in water charged with carbon dioxide.[13] The first known attempt to determine the photosynthetic quotient was made by Nicholas de Saussure and published in 1802. de Saussure had found that there was less oxygen evolved than carbon dioxide assimilated, but he also observed that an inert gas had been given off which he assumed to be nitrogen, and that the combined volume of oxygen and 'nitrogen' was equal to the volume of carbon dioxide assimilated. He suggested that the 'nitrogen' was derived from the plant tissues by decomposition[14] but Boussingault disagreed because he found that the volume of 'nitrogen' was equivalent to three times the amount of nitrogen in the plant tissues, and could not come entirely, if at all, from the plant. What was not well understood at the time was the simltaneous phenomenon of plant respiration, and although Boussingault believed that in sunlight photosynthesis alone takes place, he was aware of the large difference between the oxygen required for respiration and the much greater volume evolved in photosynthesis. He suggested in a letter to Peligot that the oxygen measured in photosynthesis was smaller than expected because of respiration needs,[15] and thus anticipated the later work of Gaston Bonnier (1853–1922) and Louis Mangin (1852–1937) who applied a respiration correction to their results by assuming, what is now connsidered to be correct, that respiration is the same during the hours of daylight as at night when no photosynthesis takes place.

Boussingault felt that an entirely new line of attack was demanded; it was not enough to know the volume of carbon dioxide assimilated and the volume of gases set free, for the problem called for a recognition that there were three systems involved, the atmosphere in the water in which the leaves are immersed for the experiment, the atmosphere in the leaves before exposure to sunlight, and the atmosphere in the leaves after exposure to sunlight. It was necessary to measure all three before conclusions could be stated about the origin of the gases liberated. In 1860 Boussingault started a marathon series of experiments,[16] forty-one altogether, using such diverse material as algae, pine-needles, the leaves of peach, laurel, water-mint, oats, holly, ivy, celery and water anemone, to name some of them. In his words, they were "experiments undertaken to find out whether nitrogen is given off by leaves during the decomposition of carbon dioxide, and the ratio that exists between the volume of carbon dioxide decomposed and that of oxygen set free". He used three large round flasks of equal size and of about a half-litre capacity which were filled with previously boiled water to which was added a quantity of water saturated with carbon dioxide gas. One flask, a control, contained water only; the second flask, also a control, contained leaves but was not

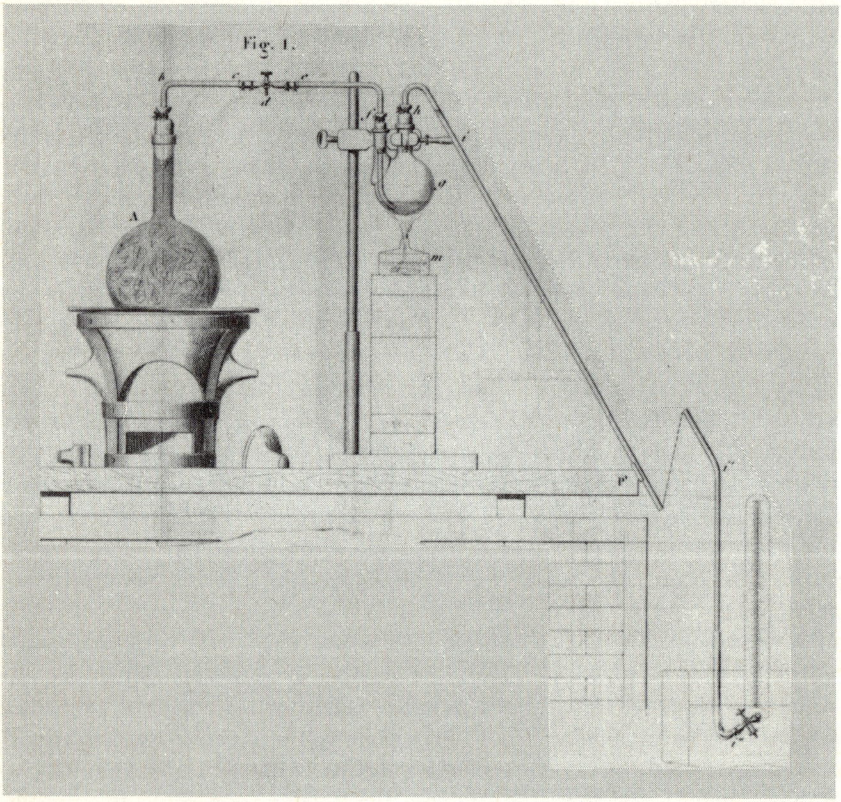

Fig. 23. The photosynthetic quotient. Apparatus used by Boussingault in 1862 to collect the gases from water in which leaves were placed whilst undergoing carbon dioxide assimilation. (*Annales de chimie et de physique*, 1862, Vol. 66, pp. 298–429 (Masson).)

exposed to sunlight. The third flask contained leaves equal in weight to those in the previous flask, but it was subjected to periods of sunlight varying from one half-bour to six hours. The gases evolved were displaced from the flasks by boiling. He found that no nitrogen was given off in photosynthesis, but more important, it was found that the volume of oxygen set free was almost equal to the volume of carbon dioxide assimilated by the leaves. de Saussure's 'nitrogen' was found to be carbon monoxide formed by the action of oxygen on the pyrogallate solution used to absorb the oxygen evolved by the leaves. Subsequent determinations using improved methods and more precise apparatus have yielded values of the photosynthetic quotient which give an average of about 1.03 although the deviations from unity are not beyond the limits of experimental error. It is, therefore, more than remarkable that Boussingault obtained with crude apparatus and bold assumptions the value of unity, a result which appears to have been generally overlooked. Surely, this investigation should be regarded as one of the classics of scientific enquiry?

The Opposing Processes of Respiration and Photosynthesis

That the subject of plant respiration made greater progress than that of plant nutrition in the first decade of the nineteenth century can be attributed to various reasons, one of which was the prevalent anthropomorphic view of nature which could accommodate the concept of respiration far better than the idea of nutrition by a gas which was a waste product of animals. Even Henri Dutrochet, on discovering that the intercellular spaces of a leaf had access to the exterior via the stomata, regarded the latter as the lungs of the plant![17] There was little awareness of the interrelationship between respiration and photosynthesis, and it was Boussingault who, in 1868, by reasoning and experiment, demonstrated that relationship in a paper on the behaviour of plants in darkness.[18] He showed that there were two opposing forces in a plant, one endeavouring to remove material and the other involving the addition of material, i.e., the opposing forces of respiration and photosynthesis, the ratio between the two depending on the temperature and the intensity of light. He germinated pairs of similar seeds, one in the light and the other in the dark, and noted the gain in weight of the former and the loss in weight in the dark – not only the weights of the plants themselves but the change in weight of carbon, hydrogen and oxygen. With a sense of wonder that never left him, he felt it strange that a plant, when dried, should weigh less than its seed, but realised that a more accurate comparison was that between the weight of the plant and the weight of its embryo. From that point he compared the egg and the seed, finding no starch in an egg, "perhaps because no-one had looked for it" [19] – how like Liebig he could be on occasion – and no cellulose in animals, except perhaps in Tunicates or sea-squirts; otherwise there was much in common between the egg and the seed.

Finally, he found that the assimilatory power of a leaf returned to normal after a prolonged period in darkness, when exposed once again to sunlight. The inhibiting factor was dessication which he believed led to the asphyxiation of the plant. If respiration ceased then so did photosynthesis. It was, perhaps, hoping for a miracle when Boussingault endeavoured unsuccessfully to revive the assimilatory property of a laurel leaf from the herbarium of Sebastian Vaillant (1669–1722) who had been Director of the Jardin des Plantes and had published a flora of the Paris region.

The Two Opposing Theories of Photosynthesis: How Does Carbon Dioxide Enter the Leaf?

Does carbon dioxide enter the leaf by diffusion over the whole of the leaf or does it enter by the stomata only? This question arose through the emergence of opposing theories – the cuticular and the stomatal theories. Boussingault had inadvertently given support to the suggestion that the gas passed into the leaf over the whole of the surface layer or cuticle. Coating leaves with vaseline to demonstrate changes in transpiration rate – or the rate at which moisture is lost – is a familiar technique in schools. With a different aim, Boussingault covered parts of leaves with black paper coated with starch paste, to measure differences in the rate of photosynthesis. Such leaves were subjected to an atmosphere of thirty per cent carbon dioxide and seventy per cent air. From his results he concluded that assimilation took place over the whole surface and not through any particular openings.[20] Support for these findings was supplied by the botanist P. Duchartre [21] who in 1866 discovered that carbon dioxide assimilation was not proportional to the number of stomata on the surface of a leaf, although he failed to state the concentration of carbon dioxide used, a factor which was later found to be of great importance. Use was made of Boussingault's results by A. Barthélemy in 1877, to establish a cuticular theory of gaseous exchange in a plant.[22]

Meanwhile, a stomatal theory of gaseous exchange had been proposed in 1865 by Julius von Sachs (1832–1897) [23] but controversy between the protagonists of the two theories was postponed until 1877–78 because of experimental difficulties and Boussingault's apparent reluctance to support a culticular theory based on his results. Because of the Franco-Prussian war, plant physiology did not figure prominently in Boussingault's work in the eighteen-seventies, and instead, he was occupied with metallurgical investigations at the Unieux steelworks of his son-in-law, Jules Holtzer. However, the various issues raised by the cuticular-stomatal controversy are mentioned here because at least two memoirs of several submitted to the Académie des Sciences were examined by a committee which included Boussingault.

The cuticular theory, as has been mentioned, received support in 1877 from a series of observations by Barthélemy – the young leaf possesses no stomata; stomatal openings are too small to account for the growth observed; the stomata provide for exhalation only because of increased gas pressures within the leaf.[24] Most of these arguments in favour of the cuticular theory were of a negarive or circumstantial kind, as also was the conclusion of E. Merget in 1878 who demonstrated that gases can enter and leave a leaf by way of the stomata.[25] Neither controversialist proved his case directly and neither made use of the fundamental proof that photosynthesis has taken place by the appearance in the chloroplasts, the green corpuscles of a plant containing chlorophyll, of minute starch granules which was the criterion introduced in 1864 by Julius von Sachs.

From Boussingault's erroneous conclusions sprang the cuticular theory but only after it was discredited were the results explained by Frost Blackman of Cambridge.[26] Blackman's method was to enclose a leaf between two glass chambers so that each face of the leaf communicated with a chamber, the two being independent. Air containing one per cent of carbon dioxide was led into each chamber and another tube connected each chamber with an apparatus for gas analysis. In 1895 Blackman found that the fixation of carbon dioxide was a function of the number and distribution of the stomata. He repeated Boussingault's experiments with oleander and obtained the same results with an atmospheric concentration of thirty per cent carbon dioxide, but when lower concentrations of this gas were used the results were reversed. The high percentage of carbon dioxide used by Boussingault had given rise to a toxic effect on the leaf cells, but also the high concentration had caused a diffusion of carbon dioxide through the cuticle which does not take place under normal atmospheric conditions.

Other Investigations on Photosynthesis

Further aspects of photosynthesis examined by Boussingault will be mentioned briefly as they are related to the technique adopted by him.[27] de Saussure had thought that leaves could decompose carbon dioxide in the absence of light[28] but Boussingault's work contradicted this claim. Nor would a plant undergo photosynthesis in diffused light, such as twilight, which surprised him as he had experienced the lush vegetation of the shady depths of tropical forests. Allied to this was a belief that photosynthesis in an aquatic plant continues in the dark after being removed from sunlight[29] but he could find no evidence to support this statement. However, since 1905 it has been known that there is a two stage concept of photosynthesis,[30] the one requiring light and the other requiring darkness during which catalytic enzymes continue the process. With a lowering of temperature to 0°C, Boussingault found that there was a decrease in the rate of photosynthesis although it was not entirely inhibited in pine needles and grass, plants typical of vegetation which survives intense cold.[31] He may not have been aware of the various factors which have produced confused results in this type of work, where the external temperature has been assumed to be that of the interior of the leaf, and where the greater viscosity of protoplasm at low temperatures might hamper carbon dioxide diffusion in a leaf.

Boussingault's researches demanded a high degree of skill in the analysis of gaseous mixtures which had been acquired many years previously in the analysis of atmospheric air at different locations. But in addition he sought a quick qualitative test to confirm that photosynthesis was taking place. This was achieved by casting small cylinders of phosphorus of various sizes which presented a variety of surface areas to an enclosed volume of gas. The plant was placed inside a bell jar resting in a dish of mercury on which a film of water was floated. The presence of oxygen in the bell jar would indicate that photosynthesis was taking place as shown by phosphorescence of the cylinder. The bell jar with plant was removed to a dark room and the cylinder, attached to a platinum wire, was inserted.[32]

Another technique was recording the colour of leaves by means of a complex colour chart, *Les Cercles Chromatiques* a monumental work of almost a thousand pages by Michel Chevreul which was particularly useful in work on pale, newly-formed leaves which were developing their chlorophyll during the experiment.[33]

One of the last of his photosynthetic investigations was about the growth of a plant started in an atmosphere free of carbon dioxide.[34] After seven weeks' growth, the plant when dried weighed less than the dry weight of the seed, and its carbon content was less showing that respiration had accounted for the loss in weight. But respiration was responsible for a little residual carbon dioxide in the container from which Boussingault concluded that the sugar and albumin which he found in the plant had been produced from the respired gas. This gave him an admirable opportunity to digress on the topic of plant and animal cells. With Dumas he exclaimed, "Animals do not create, they merely transform the materials produced by plants".[35] He was reminded of the fungi, especially those growing in the dark and bereft of chlorophyll. Pasteur had claimed that yeast and other ferments could grow in a sugar solution without the aid of chlorophyll and sunlight,[36] to which Boussingault replied that the ash of yeast contained the phosphate and tartrate of ammonia, both sources of nitrogen.[37] Again, Pasteur had demonstrated the production of acetic acid from alcohol with the aid of yeast, but Boussingault reminded him that alcohol was the product of fermentation of sugar which could not be produced in the absence of sunlight and chlorophyll. However, Boussingault did wonder whether there was a valid analogy between the growth of yeast in a chemical medium and the growth of his sunflowers of 1857 and 1858 in a sterile soil to which chemicals were added. He admitted that it was a difficult comparison for the sunflower synthesised its plant nutriments from the external world, from air and water vapour, a synthesis which a fungus cannot initiate. Although it was now 1876, Boussingault's faith in a purely chemical explanation of fermentation had not changed since 1862, which is not surprising because the complicated nutritional and chemical reactions of yeast and bacteria were unknown. A simple nutritional explanation of alcohol and acetic acid formation from sugar could not then include the unknown and undiscovered enzymes in yeast which initiated these changes. Similarly, there was no knowledge of chemosynthesis by bacteria, by which energy to synthesise carbohydrates is obtained, not from light but from the oxidation of substances in the nutrient medium.

Boussingault's achievements in the field of carbon dioxide assimilation were remarkable considering the primitive state of knowledge of plant cell structure and the complete ignorance of its mechanism. His efforts, and those of others, have been compared with a group of people examining an automobile without knowing how to raise the hood. Speed and fuel consumption could be measured, but the structure and nature of the engine causing the speed and demanding a fuel consumption woudl be unknown.[38] The complimentary contributions of photosynthesis and humus are recognised today and, as we have seen, they have had their controversial protagonists whose contributions have included the experimental work of Boussingault and the publicity given to the subject by Liebig. Yet it is Nicholas Theodore de Saussure to whom credit should be given for anticipating the contribution to plant nutrition of both humus and the products of photosynthesis.

CHAPTER 14

THE METALLURGICAL PERIOD

An event that seriously weakened the prestige of the Second Empire was the Mexican expedition of 1862. As in the case of South America the deposition of Ferdinand VII of Spain in favour of Napoleon's brother Joseph in 1808 led to a movement towards Mexican independence and, as in the South American states, there were the opposing forces of centralisation and federalism. After a shaky compromise with Spain the monarchy was finally repudiated in 1823 and a republic established on federal lines, but the introduction of central government in 1835 caused uprisings which were suppressed with the exception of the Texas settlers' revolt which led to war between Mexico and the U.S.A. and which in 1846 gave Texas, New Mexico, California, Utah and Arizona to the U.S.A. In 1861 Mexico, heavily indebted to Britain, France and Spain, suspended the payment of interest on foreign loans for two years.

In the following year Boussingault was in London for the International Exhibition of 1862 at which there were 65 classes of exhibits, each with its jury for deciding awards of merit. He was chairman of a jury for a class concerned with food substances which was subdivided into sections one of which was his special care as chairman, dealing with drysaltery or the preparation of dried and salted meat.[1] There was some confusion and misunderstanding as the exhibition organisers had asked for one juror for each of the 65 classes, instead of which the French delegation appointed 65 jurors but attached them not to all 65 classes but to those in which the French were competing and were thus loading some of the juries with their own nationals. At the last moment there was no time for reorgansiation and, as the French admitted, they found themselves a little disorganised.[2]

Whilst at this international gathering Boussingault heard that a force of British, French and Spanish troops had landed at Vera Cruz in Mexico as a debt-collecting expedition. This news was afterwards confirmed but when the British and Spanish governments learnt of Napoleon III's intention to use the expedition as an excuse to establish a French colony in Mexico they withdrew their forces and left the French to march alone to the capital, Mexico City. Through his Minister of Finance, Achille Foulds, Napoleon III summoned Boussingault to Paris to give his views on the Mexican situation, leaving immediately on May 26th with not even time to inform the French delegation that he was unable to keep a dinner date with them that evening.[3] He needn't have hurried as the Emperor did not ask for him until June 6th. Meanwhile, Boussingault was disturbed to discover that the object of the march to Mexico City was to place the Archduke Maximilian on the Mexican throne. Maximilian, nephew of Ferdinand I, emperor of Austria, was willing to accept the throne if protected by British and French troops and thus establish a French empire in the New World. Napoleon also wished to liquidate the French national debt from the output of the Mexican silver mines, but Boussingault warned the Emperor that French engineers did not understand the local Patio process by which silver was extracted from low-grade ores. He also deplored the movement of troops into the interior as they would find themselves weakened by fatigue and by lack of food and shelter, but surprisingly suggested that Mexico should be merged into the United States of America. Boussingault found the Emperior abysmally ignorant of conditions in Mexico but also a prey to delusions to such an extent that he had ordered an investigation into the possibility of transmuting metals into gold. A priest had claimed that he could change silver into gold and as a result he was granted the facilities of the laboratory of the *Comité d'artillerie* in Paris and a kilogram of pure silver from La Monnaie, the French Mint. The commandant of the artillery laboratory, Henri Caron (1823–1876) had explained to Napoleon that gold was often found mixed with silver from countries where refining was not practised, and therefore it was possible that the priest had previously obtained gold from such unrefined silver, but Napoleon was adamant and insisted that the experiments should continue, but not a particle of gold was found.[4]

Boussingault's discouragement of the Mexican campaign was undoubtedly impartial, as, at the time, the steelworks of his son-in-law, Julius Holtzer, at Unieux were showing increased profits as a result of the war.[5] The Hapsburg Maximilian became Emperor of Mexico in 1864 and at the same time Napoleon III formed *La Commission Scientifique du Mexique* to make known and exploit the natural resources.[6] With the French withdrawal from Mexico in 1867 the Commission ceased to function and only three issues of its *Archives* appeared, in which Boussingault contributed a paper on the subject of pulque, an alcoholic drink prepared by fermenting the sap of an aloe, the Maguey or *Agave americana.*[7] The analysis of aloe sap in terms of glucose, sucrose, malic acid, albumin and ammonia formed the subject of another paper.[8] On reading the three volumes of the *Archives*, one realises that he was taking an active part in the work of the Commission, for the New World and its memories were exercising a rejuvenating effect. On one occasion he wrote to his friend Dumas excusing himself from a sub-committee meeting of the Académie des Sciences as he was attending a meeting of the Commission – "As I have a report to make I cannot miss this meeting".[9]

At a dinner at the Tuileries in 1867 the Empress Eugénie asked for Boussingault's opinion of Maximilian's position. "A desperate situation," he replied, "he has been, or will be, shot". The Empress seemed not to have heard.[10] Maximilian, abandoned by his troops, was executed in June of that year.

The collapse of the *Commission Scientifique du Mexique* was not the only discouraging event of 1867 for it was a year fraught with disappointment and tragedy. Boussingault's building contractor brother-in-law, Jacques Sylvestre Vaudet, died. He was a widower, his wife, Boussingault's sister Jeannette, having died in 1859. Although she had taken a great interest in his South American adventures, perhaps a greater loss was the passing of his other brother-in-law, Louis Frédéric Achille LeBel of Bechelbronn who had so ably supported him in the field trials and animal nutrition experiments from 1836 to 1848. The bibliography of Boussingault's scientific papers shows that such large-scale experiments ceased after 1848 but that his investigations continued to necessitate access to a farm where he could examine the cultivation of tobacco, the liming of arable lands, the amount of sugar in grapes, and experiments in butter churning. In view of the excessive quantities of chemical fertiliser required today for tobacco lands, it is not surprising to read that unless a regular supply of night soil was available it was difficult to grow this crop. Boussingault recommended the establishment of tobacco plantations on the edge of towns to make use of the otherwise malodorous night soil. These were exceptional investigations for most of his papers after 1851 concern experiments which required the use of a vegetable garden rather than a farm, and this he already had at Liebfrauenberg. It has been said that Boussingault first effected the transfer of laboratory pot technique, initiated by Francis Home, to large-scale field experiments. Francis Home (1719–1815) of Edinburgh was the author of *Principles of Agriculture and Vegetation*, published in Edinburgh in 1757 and regarded as a classic among agricultural works. Yet Boussingault reverted to pot technique for many of his experiments such as those on nitrogen assimilation, and the effect of saltpetre and phosphate on plant growth.

His physical health at this time was causing some anxiety but the nature of the indisposition was unknown and could not have been serious for he had another twenty years of life. Towards the end of 1868 he apologised for his absence from a meeting of the *Conseil de Perfectionnement* at the Conservatoire because of agricultural work and the state of his eyesight.[11] The transfer of investigations from field to laboratory, and the consequent transfer of interest from agricultural topics to those purely chemical, is reflected in the change of syllabus at the Conservatoire, which, however, caused no alarm to the Administrator, Morin, though it was thought that such a change in direction might embarrass Boussingault's successor.[12]

Boussingault Moves to Unieux and Concentrates on Metallurgy

The uncertainty of the future both at Bechelbronn and Liebfrauenberg is seen in the census returns prior to 1867. Between 1836 and 1866 the population of the farm Bechelbronn fell from twenty-three to nine, whilst at Liebfrauenberg it had fallen from sixteen when the Boussingaults moved in, to three in 1866,[13] so that Liebfrauenberg was functioning mainly as a country house for the family. It was clear that for Boussingault a complete change of habitat was required and the problem was resolved by his son-in-law, Jules Holtzer, married to his elder daughter, Berthe, who invited him to work at Unieux where a new metallurgical laboratory had been built. Jules was the son of Jacob Holtzer who had established a successful steel industry at Unieux in 1835 and had been joint director of the company since his father's retirement in 1860.

Like Krupps at Essen in 1848 and Schneiders at Le Creusot in 1836, Jacob Holtzer et Cie had developed from a modest village industry.[14] Iron weapons had been manufactured at Saint-Etienne since 1664 and there is a tradition that Francis I ordered the manufacture of arquebuses, or hand guns, at Saint-Etienne in 1516. Jean Holtzer, brother of Jacob, was an inspector of weapons in Alsace where a growing timber shortage in the Vosges was compelling iron and steelworks to close and transfer their activities to the valleys where coal was obtainable for smelting in the form of coke. On being transferred to Saint-Etienne Jean Holtzer became interested in iron and steel production and induced his brother Jacob to join him. They directed a small steelworks for ten years until, in 1835, they moved to Unieux, about twenty kilometres to the south-west of Saint-Etienne where they acquired a factory for the production of wrought iron. Here, cast iron was hammered on leaving the furnace until most of the slag and impurities had been removed when a particularly tough product, wrought iron, was obtained. But the Holtzer brothers were intent on steel manufacture and in the five years after 1835 the factory changed over to the cementation process where bars of wrought iron were embedded in charcoal and strongly heated to obtain steel. As a result of this development the population of Unieux rose from 490 in 1836 to 1607 in 1848, many of the immigrants coming from Alsace owing to redundancy at the royal arms factory at Klingenthal near Strasbourg. The presence of these families was, no doubt, an added attraction to the Boussingaults when they arrived in 1867. In 1842 the Holtzer partnership was dissolved leaving the steelworks under the control of Jacob Holtzer. Today, a motorway connects Saint-Etienne and Firminy, where after several turns a minor road leads to Unieux. Opposite the main entrance to the steelworks is a narrow carriageway leading, zig-zag fashion, up a wooded hill to the Château Holtzer, built about the middle of the nineteenth century, a pleasant residence in the German style and now used for conferences. It was here that Boussingault lived and installed a laboratory after a brief spell in the bachelor quarters of the steelworks.

Nitrogen in Iron and Steel

Boussingault's metallurgical periods, apart from his assaying in South America, started in 1860 with investigations on nitrogen in cast iron and steel.[15] Such research was becoming increasingly important because steel was being manufactured by the Bessemer process which required a blast of air, as a result of which there was an increased nitrogen content with consequent brittleness. Holtzer's motives in inviting Boussingault to Unieux may not have been completely altruistic and this preliminary work on nitrogen in steel may have induced him to issue the invitation. Boussingault's sudden re-entry into the realm of metallurgy, after a lapse of thirty years, requires some explanation even allowing for his characteristic polymathic tendencies. It may have been a continuation of his absorbing interest in plant and animal nitrogen but there is another reason related to a proposed change in Conservatoire policy advocated by the Administrator, Morin, who desired a stronger link between his institution and industry. It is possible that Boussingault was encouraged by Morin to investigate the problem of nitrogen in steel because it is known that Morin directed another chemist, Jules Bouis (1822–1886), to a similar field of research in which dry hydrogen was passed over red hot iron,[16] the nitrogen of which combined with hydrogen to form ammonia which was estimated by absorption in sulphuric acid. Such a topic was well outside the syllabus of agricultural chemistry but Morin was never very concerned when professors taught outside their syllabuses provided the same topic was not treated in the same manner by two or more of them.

Whatever the reason for his renewed interest in metallurgy, the nitrogen content of iron and steel was a well discussed subject at the Académie des Sciences in 1861 through the contributions of Bouis, Boussingault, Caron, Edmund Frémy (1814–1894) and Regnault. Boussingault's first contribution to the subject was intended as a qualitative test only but he assumed that his method, the passage of steam over red-hot iron, could be used quantitatively.[17] It was a modification of an experiment by Lavoisier and Jean Baptiste Meusnier (1754–1793) in 1784 in which steam was passed over strongly heated iron to obtain hydrogen gas.[18] Nitrogen present in the steel as iron nitride united with the nascent hydrogen to form ammonia which, in Boussingault's experiment, dissolved in water formed from the unused steam. This solution was distilled and, as in his determination of ammonia in rain water, Boussingault collected the first two-fifths of the distillate which contained all the ammonia. In the face of criticism from Frémy[19] he admitted that it was not a suitable quantitative test especially as he had found that at red heat some ammonia was decomposed. This was one of three methods used by Boussingault for determining nitrogen in iron and steel.

A second method was to heat the iron specimen with cinnabar, mercuric sulphide, when mercury vapour and nitrogen were evolved leaving ferrous sulphide in the combustion tube. This method emanated from a suggestion by Regnault that a simple and suitable procedure would be to heat the iron and so expel the nitrogen which would then be collected;[20] but oxidation of the iron surface had prevented nitrogen from being driven off by heat, so instead of oxidising the iron, Boussingault found that sulphiding was a better method, hence the use of cinnabar.

Because the accuracy of these two methods had been questioned, first by Frémy and then by Caron, a third method appeared, this time a wet method. Boussingault dissolved the iron specimen in hydrochloric acid in a distilling flask, excess alkali added and ammonia distilled off in an apparatus similar to that used for finding the amount of ammonia in rain water. One of his severest critics bestowed high praise on this method, Frémy declaring that the determination of nitrogen in steel could no longer by ignored. This is still the basis of the modern method of determining nitrogen as nitride in steel to which result 0.001 per cent is added as this is the solubility of nitrogen in steel.[21]

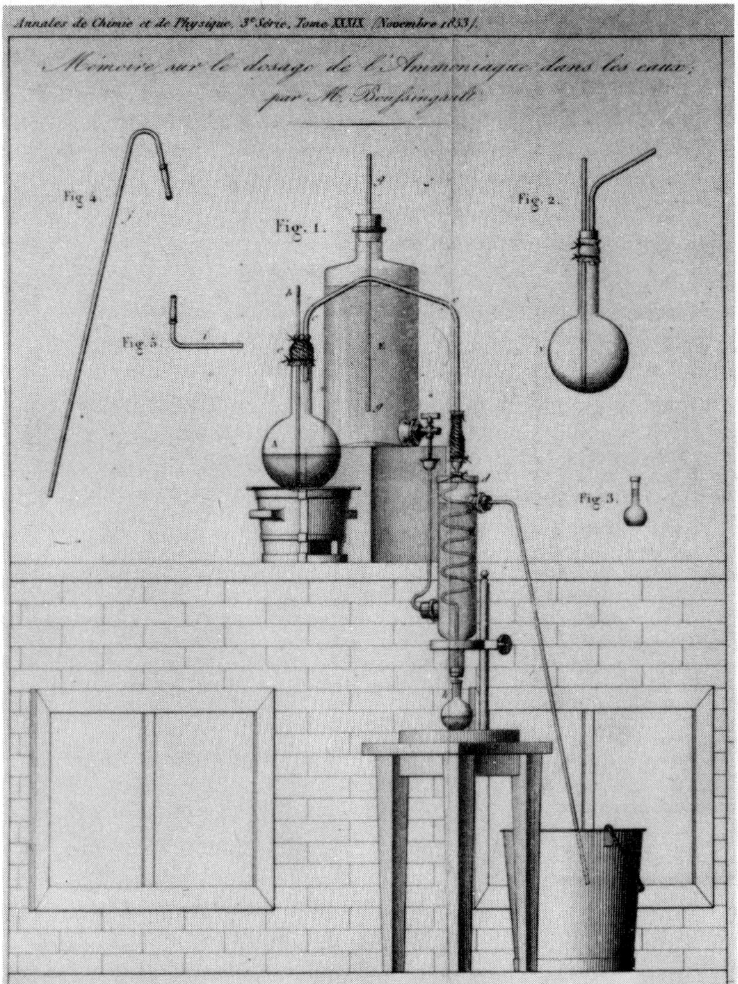

Fig. 24. Boussingault's apparatus for finding the quantity of ammonia in streams, dew, mist and rain. (*Annales de chimie et de physique*, 1853, Vol. 39, pp. 257–291 (Masson).) The apparatus was also used to find the amount of nitrogen in steel. Iron nitride in steel was decomposed by steam, forming ammonia. (*Comptes rendus*, 1861, Vol. 52, pp. 1008–1011. Archives de l'Académie des Sciences, Paris.)

The Cementation Process

His determinations of nitrogen in iron and steel were but a preliminary to his investigations at Unieux from 1867 to 1877 directed towards elucidating the steel-making process, the nature of steel, and the methods of analysing iron and steel. His first assignment at Unieux was the examination of a cast-iron sample from Medellin, Colombia, a type of cast iron which was especially useful for manufacturing the heads of stamps used in stamp-mills for crushing ore, and was, therefore, a particularly tough form of iron. Its true nature was unknown in Medellin except that it dissolved in sulphuric acid to produce a green solution, believed due to the presence of nickel but found by Boussingault to be due to chromium. It was the detailed analysis of this chrome-iron alloy[22] that marked the beginning of a series of papers on the analysis of iron and steel which occupied ten years. Boussingault acted as metallurgist to Jacob Holtzer et Cie from 1867 to 1871 when Aimée Brustlein (? – 1911) was appointed. A pioneer of the production of chrome steels,[23] Brustlien reorganised the works at Unieux and in 1874 he installed Siemens-Martin open hearth furnaces for steel manufacture. The iron and steel industry was passing through one of several periods of transformation and the need for control and understanding of processes required the services of a metallurgist. During the greater part of Boussingault's sojourn at Unieux the production of steel was by the cementation process which accounts for much of his research on this topic.

Cementation is one of the oldest methods of making steel and consists of heating wrought iron bars at about 1000°C in a charcoal mixture, the brasque, in boxes made of stone, brick or fireclay, for a period depending on the depth to which carburization is required; in the case of the Unieux furnaces the time was fifteen days after which the boxes were allowed to cool. The origins of the cementation process are difficult to trace since from the eleventh century there have been descriptions of surface hardening of tools by carburization which is the first stage in the complete conversion of iron into steel by cementation. W. Ellyott and M. Meysey patented a cementation process in England in 1614 and again in 1617, an action which aroused protests in Britain and the Continent, steel manufacturers alleging that their method did not produce steel, and two armourers of Liège claimed priority.[24] However, Robert Plot (1640–1697), the first professor of chemistry at Oxford, wrote a detailed description of the cementation process in his *Natural History of Staffordshire* published in Oxford in 1686, and wrote as if the method was new and unusual.

What substance penetrated the iron during cementation was the subject of various theories for it was not certain that the obvious element, carbon, was responsible because the composition of a brasque was usually a secret, each foreman at a furnace having his own recipe. Sulphur and volatile salts were suggested but it was Claude Berthollet, Alexis Vandermonde (1735–1796) and Gaspard Monge (1746–1818) who together in 1786 showed that the penetrating substance was carbon.[25]

Boussingault made no reference to theories of cementation of which there were two in principle, the gaseous diffusion theory and the solid diffusion theory. There were several approaches to the gaseous diffusion theory one of which was advanced by Caron in which carbon was believed to be conveyed to the interior of the iron by a volatile cyanide formed from carbon, iron and atmospheric nitrogen. On the other hand, Frémy suggested that the volatile cyanide was formed by combination of carbon and iron with the nitrogen dissolved in the iron, a variation of Caron's theory. Another gaseous diffusion theory, due to L. Gillet, assumed that carbon monoxide was formed at the surface of the iron and on penetrating it decomposed into carbon dioxide and carbon. The solid diffusion theory claimed that cementation took place by direct contact between iron and carbon without the intervention of a gas, a theory favoured by Boussingault who refers to the work of Clouet, and later Margueritte, who heated iron wire in contact with diamond dust to produce cementation steel, although there were others who had performed this experiment. The significance of such experiments was that they were carried out at a temperature below that of molten iron so that there was no possibility of gaseous diffusion. It is generally believed that carburization produced by gases entering the iron can explain cementation to a small depth but for total conversion into steel there is a molecular migration of carbon which reaches the deeper layers so that both gaseous and solid diffusion play their part.[26]

Boussingault, with a slight scorn for the theoretical aspects but with a profound respect for the empirical bases of his work, attacked two problems related to the cementation of iron.[27] Firstly, for how long did iron remain in contact with the brasque at red heat? In the absence of modern pyrometry the practice was to use trial bars, each with an end protruding from the furnace which could be withdrawn to judge the furnace temperature according to whether the bar was dark red, cherry red, bright cherry red, orange red or orange. In one particular trial he found that the iron and charcoal had been in contact at red heat for fifteen days.[28]

The second problem was the limit of carburization of iron.[29] He knew, of course, the gradations of the carbon content of iron and steel, from 0.1 per cent in wrought iron to 4.5 per cent in cast iron, with intermediate values of 0.25 per cent in mild steel and up to 1.5 per cent in high carbon steels. These values had been first determined by Torbern Bergmann (1735–1784) of Uppsala, Sweden, in 1781 but more precise figures were obtained by Berthollet, Vandermonde and Monge in 1786[30] as well as by Louis Guyton de Morveau (1737–1816) in the same year.[31] Boussingault's analyses of cast iron revealed a maximum carbon content of 4.40 per cent but by cementation he obtained a steel containing 5.07 per cent and it appeared that about five per cent of carbon was the limit of carburization made up of one per cent of combined carbon and four per cent of graphite. By combined carbon, Boussingault referred to a compound of iron and carbon, iron carbide, with a formula Fe_4C. The state of knowledge of the structure of steel may possibly be gauged by the questions posed by Boussingault,[32] for example, does carbon combine with iron only and not with any other metal in steel? Does combination with other metals account for the rise in carbon content when cementation takes place? Is graphite independent of the metallic mass? He did not succeed in answering these questions, indeed he realised the difficulty of finding such answers within the experimental limitations of the time, for he cited an example where the same steel, analysed by two workers, was said to contain 4.5 and 6.6 per cent of carbon! He saw an analogy between carbon in molten iron and common salt in a hot solution that deposits its crystals on cooling, just as graphite is deposited from a slowly cooled molten cast iron to yield grey cast iron. Yet the anomaly was that whereas he regarded the salt as in solution, and therefore a mixture, he claimed that the carbon in molten iron formed a carbide. This apparent confusion between a compound and a mixture may explain why he was almost invariably anxious to demonstrate that minerals and alloys contained their elements in atomic proportions and gave them formulae which were not justified. Had he concluded that carbon in molten iron was in solution he would have anticipated later results of other workers, for even after solidification steel is in a state of solid solution, austenite, named after a famous British metallurgist, William Roberts-Austen (1843–1902), until at a temperature between 700°C and 850°C the atomic arrangement of the iron atoms begins to change and carbon can no longer be held in solution. The average steel in greatest use contains from 0.2 to 0.3 per cent of carbon and when this solid solution cools to about 800°C carbon is precipitated as a carbide, Fe_3C, known as cementite. The iron solvent from which it is precipitated is ferrite which confers ductility on the steel and accounts for two-thirds of the mixture of which steel is composed. The precipitated cementite forms alternate layers with some of the ferrite and because of its iridescent appearance under the microscope it is known as pearlite which gives hardness and strength to steel.[33] That Boussingault did not elucidate the true nature of steel is not surprising because this varies according to the percentage of carbon. Below 0.8 per cent steel consists of pearlite and ferrite as described above, whilst that with over 0.8 per cent of carbon consists of pearlite and free cementite, a hard steel of great strength but with little ductility. As might be expected, a steel of 0.8 per cent carbon is entirely pearlite.

His Determinations of Carbon in Iron and Steel [34]

The specimens of cementation steel examined by Boussingault could not be so neatly classified as described in the previous paragraph because the wrought iron used in the process might have been produced from either white cast iron or grey cast iron according to the rate of cooling of the molten metal when it was cast from the blast furnace; rapid cooling produced white cast iron in which carbon combined with iron to form a carbide, described by Boussingault as 'combined carbon'. A slow cooling caused carbon to be precipitated as platelets of graphite. It was, therefore possible to find both 'combined carbon' and graphite in the same specimen and it was his endeavour to evolve a method of estimating both states of carbon. He described two methods – wet and dry – each depending on the chloridation of iron to ferrous chloride by the action of mercuric chloride. The carbon so liberated was burnt at dull red heat to remove 'combined carbon' and the residual graphite was burnt in a stream of oxygen. Only silica remained. Boussingault preferred the more accurate wet method but both were faulty because it was assumed that graphite did not oxidise at dull red heat, yet it was typical of Boussingault's ingenuity in devising completely original techniques when the occasion called for a new approach to an old problem. It does appear somewhat naive, if not confusing, for Boussingault to contrast his so-called chloridation method with the oxidation method of Liebig, for did not his method consist of slow oxidation to remove carbide followed by rapid oxidation at a higher temperature to remove graphite? The wet chloridation method received some recognition in one of the later editions of that standard and authoritative text of the latter half of the nineteenth century, *Quantitative Chemical Analysis* by Carl Remigius Fresenius (1818–1897) of

Wiesbaden, but Fresenius believed that, in general, methods other than that of oxidation were less trustworthy. Such a viewpoint has been more than justified since 1912 when Fritz Pregl (1859–1930) developed his methods of microanalysis at Innsbruck.

In addition to the determination of nitrogen and carbon in iron and steel, Boussingault examined methods of finding the percentage of silicon, sulphur, phosphorus and manganese but these were repetitions or variations of existing methods except the attempted electroanalysis of steel to find the percentage of manganese, an attempt which proved abortive as manganese was deposited at the cathode only with large currents. The small currents used by Boussingault yield manganese dioxide at the anode.

The dates of birth and death of Frédéric Margueritte are not known and almost nothing is known of his career except that he was employed in a Paris gasworks. In 1846 he introduced potassium permanganate solution into volumetric analysis as a means of finding the percentage of iron in an ore.[35] The principle of the method is simple; iron is converted to ferrous sulphate using sulphuric acid and then a potassium permanganate solution of known strength is delivered from a burette until the iron solution becomes pink showing that all ferrous sulphate has been oxidised to ferric sulphate. Boussingault adapted the method for finding the percentage of iron in a sample of cast iron or steel. This was a simple operation if the strength of the permanganate solution was accurately known and this was determined by using a pure solution of known strength of ferrous sulphate obtained by dissolving iron, as pure as possible in sulphuric acid. Margueritte had used piano wire but the meticulous Boussingault was not satisfied with the purity of this material so he used a method due to Henri Caron by calcining pure ferrous oxalate to the oxide and then reducing it to metallic iron with hydrogen gas. The resulting iron was dissolved in sulphuric acid to standardise his permanganate solution. His task would have been easier if he had known of the very pure ferrous ammonium sulphate or Mohr's salt. Friedrich Mohr (1806–1879) of Koblenz informed Liebig of his new salt by letter in 1855 but did not make it generally known until 1870 when he published a texbook of volumetric analysis.[36]

A Return to the Platinum Experiments of 1819

Boussingault's introduction to metallurgy occurred in 1819 at the Ecole des Mines at Saint-Etienne where he prepared a silicide of platinum and repeated

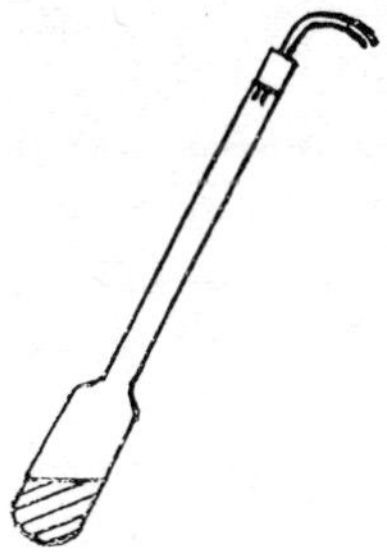

Assayer's flask containing pure iron and sulphuric acid heated in an atmosphere of hydrogen. The thin curved tube is to prevent the escape of hydrogen and the entry of air.

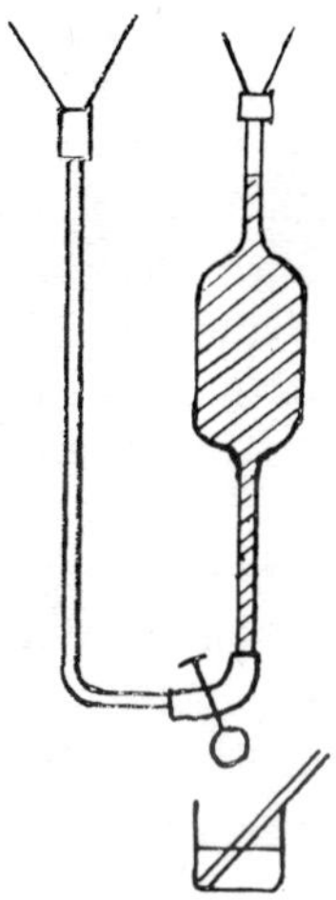

Titrimetric apparatus for filling a Stas pipette with a solution of potassium permanganate which is to be standardised by the ferrous sulphate solution in the beaker. The ferrous sulphate solution was prepared in the assayer's flask above. The Stas pipette was filled by pouring the permanganate solution into the larger funnel and opening the clip.

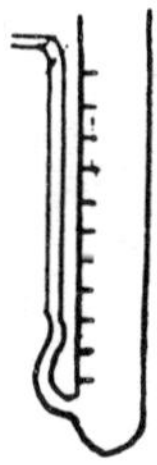

A Gay-Lussac burette. The permanganate solution was delivered a drop at a time by tilting the burette.

Fig. 25. Permanganate titration apparatus: standardisation of a potassium permanganate solution with pure iron. (*Annales de chimie et de physique*, 1875, Vol. 5, pp. 194–217 (Masson).)

Clouet's experiment on the production of an apparently carbon-free steel in which silicon had replaced carbon. He returned to this problem in 1876 when analytical techniques had improved and there was greater precision in the determination of carbon in steel. It will be recalled that he strongly heated platinum embedded in charcoal but instead of obtaining the expected carbide he obtained platinum silicide a result he ascribed to the presence of siliceous ash in the charcoal. On this later occasion he ensured that his charcoal was as pure as possible and free from silica by making use of graphite from Ceylon, sugar charcoal and charcoal from beech wood. By heating platinum in pure charcoal and obtaining only a silicide Boussingault concluded that the silicon emanated from the silica in the walls of the earthenward crucible, yet heated in a brasque of pure silica no platinum silicide was formed. Thus it appeared that carbon was an intermediary through which silicon migrated from the walls of the crucible to the platinum.[37]

With better analytical techniques he was able to show that his silcon steel of 1820, as well as Clouet's silicon steel, really did contain a small percentage of carbon supplied by furnace gases, but enough to give to the mixture the properties of steel, and so disposed of his youthful conclusion that in a steel the carbon could be replaced by silicon. Collateral proof was forthcoming when a very soft iron, with practically no carbon, was melted in an earthenware crucible and then poured to form an ingot which was found to contain 0.4 per cent carbon, sufficient to classify it as a medium carbon steel, due to the smoky furnace gases penetrating the orucible. It is not surprising that he regarded the correct interpretation of his early Saint-Etienne experiments as among the the best experimental work of his career, prompted, of course, by a certain nostalgia for those student days.

Chrome Steels

Boussingault's metallurgical analyses at Unieux had started with the examination of a chrome cast-iron from Medellin in Colombia in 1868, as already described, and ended in 1877 with an investigation of the compositon and properties of chrome steels. He presented some interesting facts about their origin.[38] Assigning priorities for the invention of chrome steel is complicated by several factors, for example, its laboratory discovery, its industrial application, and the need to distinguish between ferro-chrome and chrome steel, the former being an alloy containing about 70 per cent of chromium which is added to molten steel when a chrome steel is required. Some have given

Faraday credit for the invention of chrome steel[39] but in a paper on the subject of alloy steels Faraday and his colleague Stodart referred to "M. Berthier who first made the alloy of steel and chromium",[40] although it appears that Pierre Berthier obtained his inspiration from a previous work of Stodart and Faraday in which they recorded the results of alloying steel with platinum and two of the 'platinum metals', iridium and osmium,[41] thus confirming Boussingault's claim that Berthier was the inventor of chrome steel in 1821, a magnanimous claim since Berthier had refused to accept the results of Boussingault's early experiments on steel at Saint-Etienne. It was said of Berthier that he refused to recognise any discovery unless it had the approval of that giant of the early nineteenth century, Berzelius.[42]

Industrially, chrome steels were first produced in Pennsylvania in the early eighteen seventies[43] following a patent granted to Julius Bauer of New York in 1865.[44] Boussingault was critical of Bauer both ethically and scientifically because Bauer had not bothered to find out what previous work had been accomplished in this field; and for Bauer's belief that chromium replaced carbon in such steels.[45] Such authorities as Robert Hadfield and William Austen-Roberts agree that Brustlein, in 1877, at the Unieux works of Jacob Holtzer et Cie, was the first to produce chrome steels in Europe[46] on an industrial scale and thus enabled the upper Loire basin to become an important armaments area. Today, chrome steels contain from one to twelve per cent of chromium although stainless steels may have up to thirty per cent. As is well known, the addition of chromium to steel confers an increased hardness.

Before leaving this metallurgical period no excuse is given for mentioning his studies on the iron content of blood and body fluids of various animals using Margueritte's method which meant that the iron was dissolved in sulphuric acid and then titrated with a standard solution of permanganate of potash. Boussingault's important contribution was the discovery that the provision of iron in the diet was essential because a fully grown animal excretes iron at the same rate as its intake, although in a growing animal some iron is retained by the body. Transferring his attention to invertebrate blood he examined that of yellow slugs so common in the vegetable garden of Liebfrauenberg and was surprised to note their low iron content and the colourless elliptical corpuscles, a rare occasion when he made use of the microscope. Although there was iron in the leaves which were eaten, the slugs excreted most of it,[47] but it was not until 1885 that it was discovered that the oxygen carrier in molluscs is a copper compound, haemocyanin.

Conservatoire Criticism of Boussingault

It was about the time of Boussingault's move to Unieux in 1867 that there started a shift in emphasis in his Conservatoire syllabus from agriculture to agricultural chemistry as can be seen when his syllabus of 1862–63 is compared with that of 1873–74 (Appendices 1 and 2). He mentioned in 1868 that the mercuric chloride method of determining carbon in a steel was being taught at the Conservatoire, a topic remote from agricultural chemistry. The nature of the new course is hinted at towards the end of 1866 when Boussingault asked for an extra allocation to purchase platinum apparatus.[48] Henri Tresca, professor of mechanics, suggested at a meeting of the Conservatoire Council that a platinum model of an alembic, an ancient retort for distillation, exhibited in the galleries of the Conservatoire, could be used for making such apparatus. The project was eventually abandoned when it was found that the cost of replacing the alembic with a model in copper or tin would be equal to the value of the platinum recovered.[49] In 1869 Boussingault was criticised by his agricultural colleague, Moll, because agriculture did not occupy a sufficiently significant place in his syllabus,[50] and at the same meeting Peligot suggested that the agriculture courses be transferred to the Ecole de Grignon, near Versailles, where there were facilities for practical work. There was further criticism of the course when Boussingault requested that his two lectures per week of one hour each should be compounded into one lecture of two hours as attendance at the lectures fatigued him, especially the time spent in assembling apparatus.[51] There was criticism of the small number of students even though numbers had been augmented by a class from the Ecole Normale brought there by Pasteur. The request was not granted and Tresca reminded him that he was not indispensable, an act of discourtesy to which the members of the Council strongly reacted and assured Boussingault of its esteem for him. Official recognition of the changed character of the course did not occur until after the Franco-Prussian war of 1870–71 when at a Council meeting in July 1871 it was announced that the title of the course was now 'agricultural chemistry and analytical chemistry'. The hypercritical tendency of Council members may have been a sign of an ageing institution, for Morin and Payen were both 75, Boussingault was 68, Persoz 65, Moll 61, and of those below sixty years Peligot was 59, Tresca 56 and Laussedat a mere 51.

Boussingault's metallurgical period was something quite fortuitous, starting as it did with an opportunity to work in a new metallurgical laboratory at Unieux, and extending to the eighteen-seventies because of the France-Prussian conflict, during which, fortunately, Unieux was outside the theatre of war. The death in 1876 of son-in-law Jules Holtzer at the age of forty-two, followed by that of Boussingault's wife in 1877 determined the end of this phase, especially when his daughter, Berthe Holtzer, decided on moving to Paris. This period of bereavement and migration to a strange house in Paris had its consolation when the Royal Society in 1878 awarded him the Copley Medal "for his long, continued and important researches and discoveries in agricultural chemistry". He was in distinguished company for in the same year the Davy Medal of the Royal Society was awarded to Louis Cailletet and Raoul Pictet who, by their independent efforts in cooling the so-called permanent gases below their critical points, showed the way to the liquefaction of air and hence a cheap means of preparing oxygen on a commercial scale. The Brin process, discovered by the researches of Boussingault, was thus gradually superseded.

CHAPTER 15

THE LAST YEARS

In the shadow of an impending war, in March 1870, Boussingault made great efforts to secure for his only son, Joseph, an assayer's appointment at Besançon. Joseph Boussingault (1842–1925) graduated in 1861 and after a short period in the laboratory of Henri Sainte-Claire-Deville he became a demonstrator in December 1862 to his father at the Conservatoire and assisted him in his agricultural work in the laboratory at Liebfrauenberg. This was followed by a post as assistant to the two assayers at the Paris assay office, the *Bureau de Garantie*. The Besançon watchmaking industry, dating from the eighteenth century, was experiencing the effect of strong competition from nearby Switzerland and in 1859 the staff of the assay office there, established in 1797, was increased in order to examine the watchcases from over the border.[1] The Besançon application was the subject of a letter to Dumas who represented the *Académie des Sciences* on the Mint Committee, the *Commission des Monnaies*, and therefore could exert some influence. Wrote Boussingault, "In a long scientific career I have obtained so little for myself that there is some justice in allowing something for my son".[2] Dumas secured an audience of the Emperor for Boussingault on March 10th, 1870, but to his chagrin Dumas did not attend, no doubt tactfully absenting himself. Wrote Boussingault afterwards, "I was very disappointed in not seeing you yesterday at the Tuileries, I counted on you for the prologue".[3] If it seems incredible that the Emperor's intervention was necessary to obtain an assayer's post in a provincial town it must be remembered that government assaying was an attractive occupation financially and that there was strong competition for such posts. The French assay offices, although under government inspection, were staffed by free entrepreneurs, the assyers providing the apparatus and the reagents but were entitled to all payments made, according to a government scale of charges.[4] It was, therefore, a lucrative profession especially if one was fortunate to be appointed to a large city or a town associated with jewellery or watches. In Paris these posts had attracted eminent scientists and Dumas had personally intervened to obtain such posts for deserving scientists,[5] and no doubt his influence secured the Besançon appointment for Joseph in May 1870, an appointment which kept him away from the war zone of 1870–71.

Like his father, Joseph had a tremendous flair for analysis. In 1878 he was given the repetitive task of analysing the wines, over ten thousand of them, which were submitted for the Exposition Universelle in Paris of that year. Not only did he complete this formidable assignment by evolving a new technique of rapid fermentation but also earned the praise of no less an authority than Pasteur.[6] From 1880 to 1889 he was an assayer at the Paris Mint and then at the Assay Office where his reports can still be read. Joseph appears to have been the most esteemed and popular member of the Boussingaults. His love of the Alsace countryside is reflected in his photographs some of which were issued as picture postcards. We have seen the effect of the intimate professional relationship between father and son in protesting about the copper content of kirsch, a subject somewhat related to Joseph's investigations on the fermentation of stone fruits. In his anxiety to promote his son's career and in his involvement in Joseph's researches, Boussingault was apt to take hasty action in correspondence which was, for him, completely out of character.

The Franco-Prussian War of 1870–71

The growing influence of Prussia in Europe was a matter for concern with the French government. In its ambition to become the major power in the confederation of German states in 1866 Prussia had waged a successful war against its rival Austria culminating in the victory of Sadowa. To obtain an ally in Spain, the sister of Isabella II was married to the Duc de Montpensier in order to secure a French heir to the Spanish throne. Isabella's husband was thought to be unable to father an heir, but French hopes were dashed when Isabella gave birth to a son, and further dashed when the monarchy was overthrown by General Prim's coup d'état of 1868. The throne was offered to various royal families, all of whom were unwilling to assume the title, Queen Victoria actually forbidding any member of her family to accept. But a Prussian prince, a Hohenzollern, seemed attracted – he afterwards withdrew – and this Prussian interest in Spain caused immense alarm in France. The French government, far from satisfied with the Hohenzollern withdrawal, insisted on a personal guarantee from William I of Prussia that no Hohenzollern would occupy the Spanish throne, who, of course, could not grant it. The provocative reply, the famous Ems telegram inspired by Bismarck, led to a declaration of war on July 19th, 1870 by the Imperial Council, now directed by the Empress Eugénie as Napoleon was too ill.

An intial French victory outside Saarbrucken was followed by a Prussian invasion when the Third Army Corps crossed the Alsace border north of Wissembourg on August 4th, the town falling at midday. Liebfrauenberg, 15 kilometres away, was in the path of the advancing troops and there is a tradition that they bivouacked in the woods of Bechelbronn that night whilst their officers were accomodated at Liebfrauenberg.[7] This may be partially true but it is unlikely that officers in time of war would leave their men whilst they found shelter five kilometres away. So the thermometer, taken, with apologies, as a memento by a German officer may have belonged to Boussingault or to young Joseph LeBel, the chemist, who had a laboratory at Bechelbronn. The same officer had told his fellow officers to leave the house undamaged as it was of scientific importance, contrasting favourably with the treatment meted out to Regnault whose laboratory at Sévres was wrecked and his papers and books stolen.[8] However, the wine at Liebfrauenberg, intended for the French soldiers, was commandeered by the invaders together with oats, wheat, cows and horses from Bechelbronn.[9]

August 6th saw the bloody battle of Froeschwiller, near Liebfrauenberg, where both sides each lost ten thousand men in killed and wounded, the local population taking a week to bury the dead. Boussingault did not entirely escape damage to one of his properties. He was absent from a Conservatoire meeting in October 1871 because of repairs to a château at Reischoffen, nine kilometres to the south-west of Liebfrauenberg.[10] It was here that a detachment of French cavalry, or *cuirassiers*, was annihilated in attempting to charge the Prussians who fired at them from vantage points in the village houses. In the new form of warfare, as understood by the Germans, but not by the French, heroic gestures were not only unsuccessful but were suicidal.

Since 1818 the French army had been recruited by a ballot system. If you were in an age group liable for military service you might draw a 'bad number' in which case you entered the regular army, or a 'good number' which rendered you liable for occasional training in the reserve. But exemptions could be purchased and agencies existed for this purpose – 'that poverty should pay this blood tax' as was said of the substitutes who volunteered because they were out-of-work. In 1868 was formed the Garde Nationale Mobile for those who had escaped the call-up or had purchased exemption but when war was declared it consisted of half a million men who were unorganised, unequipped and untrained.[11] Joseph Boussingault was liable for service with the Garde Nationale Mobile but regulations were unnecessarily rigid for such a time of crisis because he could only enlist at his former place of residence

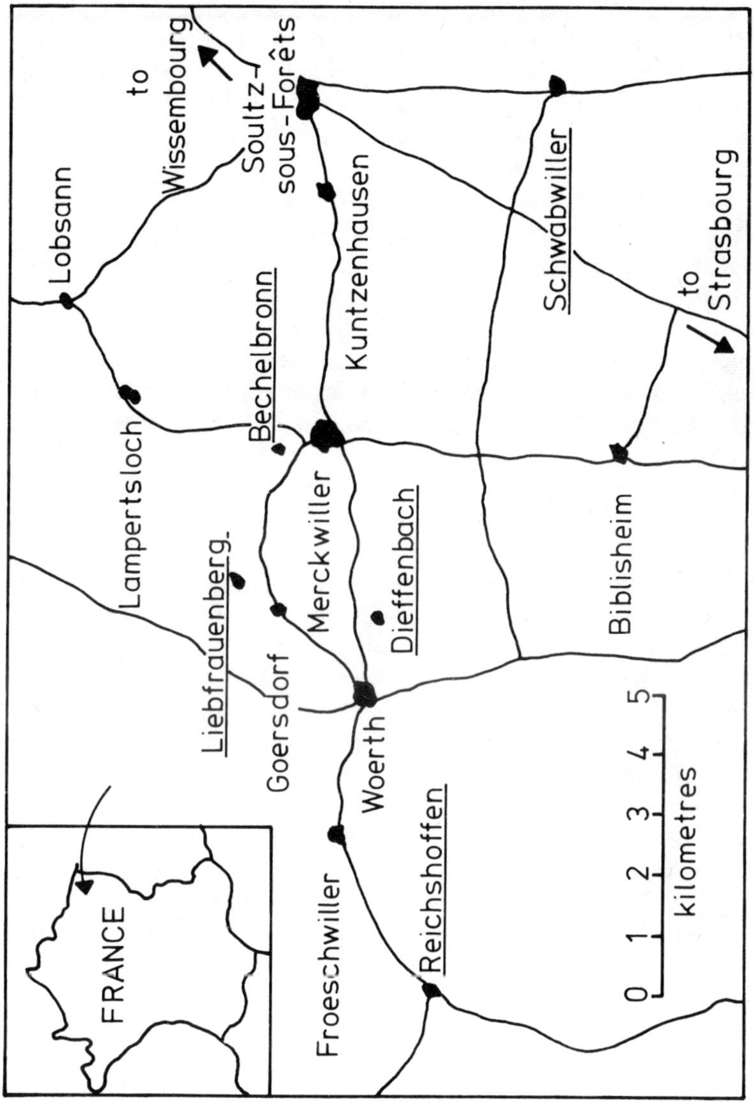

Fig. 26. The Alsace of Boussingault and the LeBels. (The family estates are underlined.)

which was either Paris or Strasbourg and not at Besançon where he was now living as he was a newcomer to that town.[12]

Boussingault was then living at Unieux and was therefore safely out of the fighting zone but news of Bechelbronn and Liebfrauenberg was sent by those in the district who either corresponded with him or his daughter Alice, married to an ironmaster, Jean Claud Crozet (1831–1892) at Chambon Feugerolles, between Lyon and Saint-Etienne. Crozet was also a Conseiller-Général, and elected representative to the administrative council of the Loire département, similar to a county council in Britain. Joseph LeBel, the future pioneer of stereoisomerism, then 23, reported on having gone to Froeschwiller to purchase horses to replace those that were commandeered from Bechelbronn by the invading forces. He returned with the news that cholera was rife due to "the odour of corpses buried in shallow graves", a tribute to the persistence of the miasmatic theory of disease. Boussingault was relieved to hear that the fabric of Bechelbronn and Liebfrauenberg had been spared, especially was he concerned about the fate of the fourteenth century tower. Few Prussians had passed that way after the beginning of August and those had been accommodated not in the old monastery but in the former pilgrims' hostel, Greyenbuhl.[13]

A studied idiosyncrasy in Boussingault's correspondence with son Joseph was the dating of letters as '1815' instead of 1870, as a reminder of a previous year of disaster which had seen the entry of foreign troops into Paris, the second Restoration of Louis XVIII and the White Terror when those who had favoured Napoleon were hunted and executed. The chaotic state of military administration was shown by the demand of the conscripts at Unieux for the minimum regulation period of training normally given to army personnel before going into the field – an inadequate fifteen days. There was also the possibility of a call-up of older men for we hear of Boussingault practising with a rifle and wryly remarking "that it will not be agricultural to plant a bullet in the back of a Prussian".[14] He writes to Joseph complaining that the lack of war news is having an unsettling effect on the people of the Loire around Unieux, and on one occasion the Préfet had called on the firemen, a para-military body, to prevent an angry crowd from marching on his office.[15]

Dislocation of the work of institutions and societies was inevitable, especially during the siege of Paris. The Conservatoire still functioned but public courses had ceased. A committee of the engineering staff, *Commission du gênie civil*, met under the chairmanship of Tresca the Conservatoire professor of mechanics, to advise the government on the manufacture in Paris of an early design of machine gun, the *mitrailleuse.*[16] Research on dynamite was carried out by P. Champion a chemistry demonstrator.[17] Milk was manufactured during the siege on the basis of various analyses including that of Boussingault but the results were not very satisfactory.[18]

The siege of Paris started half-way through September 1870 and the first meeting of the Conservatoire after the invasion was poorly attended. "Messrs Boussingault, Moll, Peligot and Becquerel are absent from Paris because of circumstances outside their control. Some have been captured and held during the siege", so run the minutes of 15th October. Peligot was at Bordeaux to organize a mint for the production of money in that town. The terms of the 1871 settlement are well known; the cession to Germany of the southern portion of Lorraine, all of Alsace except Belfort, and the payment of an indemnity of one thousand million francs, or forty million pounds sterling, which was raised by loans but voluntary contributions were also made. The professors of the Conservatoire agreed to contribute one-tenth of their salaries "for a national subscription to liberate their country".[19] A body not so fortunate was the *Société Centrale d'Agriculture* to which Boussingault had complained that he had not been named a Correspondent of the Society for some time. The secretary replied that he was gathering all such information for the new year-book or calendar but owing to the dislocation of postal services he did not even know the names of those who had died![20]

Aimé Laussedat has been mentioned as a professor at the Conservatoire and he became the Administrator in 1881. He tells of a meeting with Boussingault whilst he was superintending the fixing of the new boundary stones of Alsace. Said Boussingault tearfully, "If this had happened in Humboldt's time he would have died of shame for the duplicity and brutality of his country's statesmen and the stupidity of those of our country which he loved as much as his own. Endeavour, my friend, to live long enough to pull up the stones which you have set up".[21] But Laussedat, although living to be 88, died in 1907, eleven years before Alsace-Lorraine was restored to France. Altogether, the transfer to Alsace to Germany had a profound emotional impact on Boussingault.

Return to Liebfrauenberg

Visits to Liebfrauenberg recommenced in 1871, and in 1872 he wrote that he had noted very little 'Prussianism' in Alsace but he was a witness of the administration's thoroughness. He described how once a week, in each canton, there is a public lesson in agriculture given by an itinerant teacher where all the leading citizens, mayors, deputy-mayors, priests, pastors and civil servants were compelled to attend, for "here one obeys and the attendance is large".[22] He describes such a lesson at Woerth, about four kilometres from Liebfrauenberg, and socially it must have been a pleasant occasion for he mentions that he met all the old folk from Liebfrauenberg. Yet professionally it was a sad day for him as the teacher lavished praise on the owner of the farm where the lesson was held for his agricultural results which, many years previously, had been attained by Boussingault himself. He describes the day in a letter to Dumas in which, for the first time, he claims that Bechelbronn was the first agricultural experimental station. Consolation, however, was forthcoming in the presentation in 1872 of the Thenard medal by the *Société d'Encouragement pour l'Industrie Nationale* for his services to agriculture over a long period.[23]

He Complains to Dumas About Assay Charges and Abolition of Posts

Whether or not it was the aftermath of war or advancing age, Boussingault became highly critical of policies adopted by the Mint. These were personal criticisms as they concerned the work and progress of son Joseph. Boussingault's health had always been remarkably good and only once did Mme Boussingault mention a slight indisposition when 'J.B.B.' had eaten too many green peas and was now on a diet. This was in 1858 when a dinner date with his former class-mate of Saint-Etienne days, Benoît Fourneyron, had to be cancelled.[24] The winter climate of Paris was an excuse in applying for Joseph's transfer from Besançon to Paris. "I am not like you, perpetually young", he wrote to Dumas in September 1875 from Liebfrauenberg, "I am getting old and there is nothing wrong in having my son near me during the winter".[25] Another factor may have been the deteriorating health of Mme Boussingault who had become diabetic.[26] He lodged a formal complaint about the assay charges in Paris even though the chief assayer was his Conservatoire colleague Peligot. His displeasure with the Paris office was due to the lower fees charged by their assayers as compared with those of the Besançon office. Tactless, perhaps, with a son in the assay office at Besançon, but Dumas, with his usual sense of correct procedure, passed on the complaint to the Director of the Mint who admitted that Peligot's charges were in error but that Boussingault had been too severe in his condemnation of such a slight difference in the assay charges.[27]

Unaffected by this rebuff over the assay charges, he objected in the following year to a government proposal to abolish the salaried posts of assay inspectors, Joseph being a prospective candidate.[28] The strain resulting from these altercations induced him to spend the winter at Liebfrauenberg instead of Paris, but not idly. In a letter to Dumas he mentions his laboratory work and a literary task engaging in a history of the conquest of New Granada which became Colombia, and the discovery of emeralds and diamonds. This history was never published but formed part of his five volumes of *Mémoires.*[29]

The affair of the assay charges, which led to a court case not involving Boussingault, may have determined the government to appoint a *Commission de Contrôle de la Circulation Monétaire* in 1879 on which Dumas and Boussingault represented the *Académie des Sciences*, Dumas being chairman.[30] One of its functions was to keep coin specimens in a locked box until the end of the year when they were assayed, as in the trial of the Pyx in Britain. In 1881 Boussingault reported on the assays of bronze five and ten centime pieces which he had supervised, but the analyses had been made by "a former assayer of the Besançon assay office".[31] Joseph was now in Paris.

Widower Boussingault Returns to Paris and Enquires About the Rollet Process

Adèle Boussingault died on June 6th, 1877 according to the inscription on the family grave in the cemetery of Père Lachaise in Paris. Little is known about her except that she was a very good wife and mother. No detailed account of her survives which is so evocative as that by Boussingault in his *Mémoires* when recollecting her as a small girl of five, a wild little open-air girl, sunburnt, with flaxen hair, and dressed in a petticoat of coarse material. Some letters to her son Joseph help to fill the gap. She was interested in her family, especially her elder daughter, Berthe, and her children. A letter from Adèle is written during a music lesson given by Berthe to her children who taught them a typical young ladies' curriculum consisting of singing and violin, drawing, gymnastics and horse-riding.[32]

It has already been mentioned that Mme Berthe Holtzer and Boussingault left Unieux for Paris after Adèle's death, and lived in Mme Holtzer's newly acquired house, 49 rue d'Anjou, near the church of the Madeleine, where he dictated his *Mémoires* to Berthe which were published in five volumes from 1892 to 1903 in a limited edition of three hundred copies. The *Mémoires* cover the first thirty years of his life, from first recollections of Paris to his return from South America. Other members of the family expedited the publication as Berthe died in 1890. On losing her husband, Jules, in 1876 she perhaps desired a complete change from Unieux, or as she was musical the city life attracted her, especially as it is known from a letter of Mme Boussingault that Berthe was bored when staying in Alsace. Although far from Unieux, Mme Holtzer maintained a lively correspondence with Aimé Brustlein who was now technical director of Jacob Holtzer et Cie, and Augustus Kopp the administrative director. What is surprising for that period is that on her husband's death she succeeded him as a co-director for two years, and therefore it is not at all astonishing that the replies of the directors to her letters indicate that she had a detailed knowledge of the technical and financial minutiae of the firm and also of the Holtzer investments.[33]

The most remarkable reply is from Brustlein, seven foolscap pages from one who excuses his late reply because for several days he has been supervising the pouring of cannon from 6.00 a.m. to 11.00 p.m.! It is dated 17th November and was probably written about 1880 in reply to a query raised by Boussingault on the proposed 'Rollet' process by which steels containing a high percentage of phosphorus could be dephosphorised. The subject was of prime importance to Brustlein because he had to decide whether to install basic linings to his Siemens-Martin open hearth furnaces, constructed in 1874, or to use the new process of M. Rollet, a dilemma which was ever present with nineteenth century ironmasters. Up to the middle of the century steel was in short supply because of its slow method of manufacture by cementation, until various inventions led to the Bessemer converter in 1856. Shaped like a large concrete mixer, the Bessemer converter is supplied with molten pig-iron from a blast furnace and the impurities oxidised to a slag by a blast of air. The Siemens-Martin, or open hearth process was introduced in England by Charles and Frederick Siemens, and in France by the brothers Martin in 1868. The molten metal lies in a shallow pool on the bottom of the furnace or hearth. Iron is heated by a gaseous fuel, producer gas, a mixture of carbon monoxide and nitrogen obtained by passing air over heated coke, the fuel being preheated by the outgoing hot gases. It is costlier compared with the Bessemer process because an external fuel has to be supplied, but there is a higher yield of steel and cheap steel scrap can be used. The hearth was originally lined with silica firebricks which removed carbon, silicon and manganese from the iron, but pig-iron containing a high proportion of phosphorus could not be used successfully in an 'acid' or silica firebrick hearth.

This problem had already been solved for the Bessemer process by Gilchrist and Thomas in 1876 by using a 'basic' lining of of limestone or dolomite. Similarly, in 1883, Gilchrist and Thomas lined the open-hearth furnace with a basic lining. Lime was also added to the molten iron with the result that one-quarter of the yield was 'basic slag' or calcium phosphate which has a ready sale as an agricultural fertiliser.

This was the stage reached by Brustlein about 1880 when British and German steels were ocmpeting too successfully with French steel which had too high a phosphorus content, and were consequently brittle, owing to the high phosphate content of French iron ores. Brustlein endeavoured to spy on a steelworks in Austria when a delegate to an iron and steel congress but he was discovered by the management, although not before he had satisfied himself that the open hearth furnaces were fed with low phosphate ores which was the reason for their superior steels; there was no attempt to dephosphorise.[34] Meanwhile, various suggestions were being made and processes tried, such as the 'Rollet' process, the subject of Brustlein's reply to Berthe Holtzer, M. Rollet being an employee of the Saint-Chamond steelworks, on the road between Saint-Etienne and Lyon. A very serious and worried man, he was willing to come to a financial arrangement with the Unieux works although Saint-Chamond would require one-third of the royalties, the process having been developed at their works. The method was to remelt the cast iron from the blast furnace in a foundry furnace adding a mixture of limestone and fluorspar, either of which would convert phosphorus into basic slag. There is no evidence that this process was ever adopted owing to the introduction of basic linings to the open-hearth furnaces.

Boussingault's enquiry about the 'Rollet' process reminds us that his metallurgical interests were still very much alive when he was in his late seventies, and even in his early eighties; in 1884, he was unwillingly drawn into rebutting some industrial and commercial blackmailing by a rival firm. Jacob Holtzer et Cie owned a steelworks at Ria in the Pyrénées Orientales, near the Spanish border, which was regularly visited by directors from Unieux. Rumours were spread that the Ria castings were likely to be brittle owing to the alleged high phosphorus content of the steel. Correspondence between the Unieux steelworks and Mme Holtzer in Paris expressed concern over the possible effects of these rumours and emphasised that all would be well if only Boussingault would make an analysis of the Ria steels. As a relation of the Holtzers he probably realised that his analyses would be suspect and he was not at all prompt in conducting them until 1884 when he agreed to execute the work and found that the Ria steels were purer than those previously produced there, especially with regard to phosphorus content, and that therefore there was no foundation for the rumours.[35]

The End and the Fate of Bechelbronn and Liebfrauenberg

Jean Baptiste Boussingault died at his daughter Berthe Holtzer's house on May 11th, 1887 in his 86th year. The funeral service,[36] which was private, took place on May 14th at the Temple de l'Oratoire, a Protestant church in the nearby rue Saint-Honoré, and after the service there was a long cortège to the cemetery of Père Lachaise where the family grave is surmounted by a bust of Boussingault. As a former colonel of the Colombian forces of Bolivar he was given a military funeral with an impressive cavalcade consisting of the 124th infantry regiment, a squadron of the 6th cuirassiers and a section of the 31st artillery. Wreaths came from the Conservatoire, students of the Institut Agronomique, Ecole des Mines de Saint-Etienne, and the distinguished pall-bearers represented the Conservatoire, the Académie des Sciences, the Prefecture, the Assemblée Nationale and agriculture, among them being Schloesing, his successor at the Conservatoire, and Laussedat, the Administrator of the Conservatoire.

Bechelbronn, which had been managed since 1882 by Joseph Achille LeBel the chemist, was sold in 1888[37] to an Alsatian firm then under German control, the *Pechelbronner Oelbergwerke*, and the farm was let to a family who worked it until 1913. With the recovery of Alsace and Lorraine from Germany in 1918, the oilfield and farm were acquired by the *Pechelbronn Société Anonyme d'Exploitions Minières* which converted two portions of the farm into housing estates for their staff and workers, the cité LeBel on the Lobsann road for staff, and the cité Boussingault on the Woerth road for the workers. During World War II the château LeBel was an administrative centre for the German occupying forces; today it stands in very good condition and is well kept by the present owners, M. and Mme Gilbert Kapp.

Liebfrauenberg became the holiday home of the Boussingault family and was inherited by the three children, Joseph, Berthe and Alice. After World war I there was considerable delay on the part of the German authorities in handing back the old monastery and its ground to the family and this led to a lawsuit against the German state in February 1924 when the family was awarded about ten thousand francs and five per cent interest as from November 4th, 1918.[38] The lawsuit was instigated by Joseph, Alice, and Berthe's daughter, Mme Cécile Mourier, a widow. In 1954 Mme Mourier,

Fig. 27. The grave of Boussingault in the cemetery of Père Lachaise. (Photograph: F. W. J. McC.)

Fig. 28. Château LeBel in 1938. On the right is the building that housed the laboratory of Joseph Achille LeBel, now a restaurant. (Photograph presented to the author by M. and Mme G. Kapp.)

the sole inheritor, sold Liebfrauenberg to the Lutheran Church of Alsace and Lorraine and after interior alterations had been made which eliminated Boussingault's laboratory and library, it was opened in 1960 as a residential centre for the Protestant youth of Alsace,[39] the chapel being rededicated in 1965, and thus the old monastery continues as a place of peace, enlightenment and inspiration. Standing on the hill overlooking the village of Goersdorf and the foothills of the Vosges beyond, one feels that here are the ideal surroundings for reflection and meditation.

Fig. 29. Liebfrauenberg before renovation. (From *J. B. Boussingault à la Société Centrale d'Agriculture*, by M. Lenglen (Imprimerie Centrale Administrative, Beauvais).)

Fig. 30. Liebfrauenberg today. (Photograph: F. W. J. McC.)

CHAPTER 16

ASSESSMENT

What manner of scientist would he have become had he not married Adéle LeBel? We can dismiss any thought of him returning to South America, even if he had married Manuelita Paris, the emerald heiress, in 1825 for the engagement, if it could be called that, was broken off because of her refusal to live in Europe, whilst Boussingault was anxious to return. Related to this theme is the seriousness of Humboldt's wish to dwell in Mexico or South America. He lived to be almost ninety and was free to return if he had felt an urge to migrate. Would he have returned if Boussingault and Manuelita had married and resided in Colombia? Pepé Paris, her father, owned farms and, like the LeBels, could have given him one of them for the establishment of an agricultural research station, but was there the professional opportunity such as was provided by a chair at the Conservatoire? More likely, Boussingault would have accepted financial security as a geologist or an assayer for a mining company. Or had he returned alone to Europe without marrying into a farming family he might have been satisfied with the posts of professor and dean of the faculty of sciences at Lyon. But a farm and professional prestige were necessary to him and these were provided by Adèle LeBel and the Conservatoire, but especially Adèle because the professorship could only have been modestly rewarded for two lessons a week for half the year; also the farm Bechelbronn was not a paying concern. Boussingault's career problem was undoubtedly solved by the oil of Pechelbronn which maintained him and his ménage, and granted him the leisure to conduct agricultural research. Meanwhile, the Conservatoire provided a platform from which he could expound the agricultural practice based on his Bechelbronn experimental trials.

The Conservatoire – An Assessment

An assessment of Boussingault requires a critical survey of the Conservatoire, especially the agricultural course. The Conservatoire developed as a school outside the faculty system of the Sorbonne and the *grandes écoles* such as l'Ecole Polytechnique and l'Ecole Normal Supérieure, from humble beginnings as a school of industrial arts and orafts. In some respects its teaching may have been of university standard but it provided continuation courses only, without examinations or certificates. There was no precise separation between the functions of the university and a quasi-technical college like the Conservatoire – certainly not the clear distinction that existed in Germany between the university and the *technische hochschule.*[1] In the somewhat vague category in which the Conservatoire was placed it would seem that a two-hours per week professorship in the winter months only was far less prestigious than the post of Dean of the Science faculty at Lyon, but of course the Conservatoire was in Paris with all its charismatic glamour and opportunities. It was essential for Boussingault to capture a base in Paris to ensure the dissemination of his researches at Bechelbronn not only through his authority as a professor of agriculture but through the medium of journals with which he was closely involved.

Yet the agricultural course could hardly be called impressive. The students hailed from farms and, judging by the syllabuses, they could not have learnt much about agricultural techniques. Instead, they appear to have heard about the crop and nutrition experiments which Boussingault was conducting at Bechelbronn. The future careers of the students were no doubt assured by inheritance of the farms on which they worked as they were the sons of farmers. There would be no need to suggest or create new careers for the students as happened at the Ecole Normale Supérieure under Pasteur.[2]

An agricultural course without land to farm is an anomaly which was hardly justified even if you held, as Boussingault did, that theory and practice could not be taught together. Louis Moll (1809–1880) was the other professor of agriculture and taught practical farming but he could not, it seems, protest about the absence of farming land. Yet Boussingault's former adversary, Georges Ville, whilst teaching at the Muséum d'Histoire naturelle, organised fertiliser trials in the park at Vincennes not far from the Conservatoire.[3] Nor is there any evidence that Boussingault ever suggested problems to be elucidated on the students' farms. At a time of development in vocational education in France it does appear that the department of agriculture at the Conservatoire was in a backwater, away from the mainstream of agricultural education, interesting and often original, but still a backwater. Perhaps the Minister of Commerce and his department should have handed over the responsibility for the Conservatoire to the Ministry of Public Instruction?

Fortunately, the professors were not compelled to teach to strictly prescibed syllabuses nor were their research projects dictated by a governing body as was the case with the Royal Institution of London in the early nineteenth century when, for example, Davy had to relinquish his interest in galvanism to carry out research on tanning processes [4] and soil analyses [5] for the influential landowners who were well represented on the governing body. Exceptions can be found when Morin formed a committee charged with the investigation of the potato disease, or directed his chemists to determine the nitrogen content of iron from the blast furnace.

Boussingault's various interests might render an adequate appraisal difficult to achieve, but in spite of his participation in several scientific disciplines an assessment is attempted under three categories, as a man, as a farmer, and most important of all, as a scientist. It is as a scientist that the most significant assessment should be made in view of his diminished reputation, not only internationally but in his own country.

The Man

What kind of man was Boussingault? The testimony of his family is vague, two of the great-grandchildren giving the impression that he was authoritarian but kindly, especially towards his wife and family. The testimony of his contemporaries is confusing for their evidence was expressed after his death and much of it appears to have been inspired by a reading of his *Mémoires*. Much of his correspondence suggests that the *Mémoires* portray a character which Boussingault would have wished to assume, bold, decisive and friendly, as contrasted with the diffident, anxiety-ridden and solitary Boussingault of his letters. Achille Müntz, his former demonstrator, perhaps came closest to the truth when he remarked that Boussingault allowed only a few trusted scientists in his laboratory and preferred, with the exception of his demonstrator, to work alone.[6] Müntz was speaking at the unveiling of a monument of Boussingault by the famous sculptor Jules Dalou in the grounds of the Conservatoire in 1895. Apart from a natural introvertness, his solitariness was also the result of suspicion, especially with regard to Liebig and his assistants at Giessen, coupled with a scepticism of other workers' results.

Boussingault left school at twelve years and in the following four years he did as he pleased, so that it was possibly a traumatic experience to submit to the discipline and scholastic cooperation of the Ecole des Mines of Saint-Etienne. This difficulty may have been partially overcome by his appointment as student-demonstrator, a post which would have given him sufficient status to avoid student friendships, yet becoming acceptable to the staff; not a situation in which he could develop a desire to cooperate with his peers.

An attempt to evaluate Boussingault's personality by a Soviet agriculturist in 1936 used Liebig as a yardstick.[7] The method used was the rigid classification of scientific personality expounded by Wilhelm Ostwald (1853–1932) in his volume on great men, *Grosse Männer,*[8] by which there was a division into Romanticists such as Davy and Classicists such as Faraday, the extremes in a spectrum of personality and as limited a division as the demarcation of extroverts from introverts. Ostwald related his division of personalities to the four Hippocratic humours, blood, phlegm, black bile and yellow bile, an excess of any one of them producing different temperaments, the sanguine, the phlegmatic, the melancholic and the choleric. Romaticists were sanguine or choleric whilst the Classicists were phlegmatic or melancholic. The Soviet writer extended this thesis by comparing the romanticist Liebig with the classicist Boussingault, contrasting the basic experimental work of Boussingault with the theoretical deductions in agriculture of Liebig who greatly relied on the observations of others on which to construct his hypotheses. Boussingault examined every possible objection to his experimental conclusions. The Romanticist in Liebig was demonstrated by his popularisation of agricultural science by means of his books, written in a polemical style in which he even threatened farmers with a smaller harvest if they did not heed his warnings.

The weakness of the Soviet thesis follows from that of Ostwald with the limitations due to extreme cases. The contrasting characteristics of Boussingault and Liebig are to a certain extent valid, but the generalisations are too sweeping. Liebig did conduct some agricultural trials even if they were supervised by others, an outstanding example being the fertiliser trials which were conducted in five different parts of Saxony from 1851 to 1854. Again, his popularisation of agricultural science projected his provocative ideas before a much larger audience than the readership of Boussingault's *Economie rurale*.

If we move away from this limited assessment and endeavour not to fit Boussingault into a preconceived psychological category but to assign him to one of the many personality groupings, then a more satisfactory result is obtained.[9] It may be agreed that he was somewhat seclusive and non-social and therefore inclined to avoid not only competition but cooperation. He found so much satisfaction within himself that he tended to be emotionally and intellectually independent of others, a characteristic which rendered him incapable of close relationships but left him aloof and emotionally detached. His personality might be described, therefore, as mildly schizoid but very far from the extreme state of paranoia such was exhibited by Georges Ville.

There were, however, occasional paranoiac phases which needed well-intentioned rebukes from Humboldt and brother-in-law Sylvestre Vaudet. Boussingault's unsuccessful attempts to become an eminent citizen through political rather than scientific advancement may have been paranoiac phases or may have merely been a wish to emulate his close friend J. B. Dumas who, until the coup of 1851, had successfully pursued a parliamentary career and thereafter had served on many committees of national importance. The unsuccessful struggle for status can leave in its train a sense of inferiority and this is well illustrated in his letters to Dumas. His type of personality may have accounted for breaking association with Benoist at Saint-Etienne, with Rivero in South America and with Ville at the Conservatoire.

This brief analysis may explain why Boussingault founded no research school although the enviornment was propitious as he was directing an agricultural research station from the age of thirty-nine and had been conducting research on the farm Bechelbronn for the previous five years.Liebfrauenberg could have accommodated a school of research workers which, with the experimental techniques developed by Boussingault, might have extended his investigations over a wider area and to a greater depth. Yet we hear of no assistant at Liebfrauenberg and Bechelbronn apart from his son Joseph and the rarely mentioned Letellier who performed the analyses of soils and crops in the five year rotation of 1836–41. His seclusive and detached manner was utterly inconsistent with the development and maintenance of a research school for he lacked the charismatic power to attract and hold students even though one may be sceptical about the possibility of a regular supply of such research students, for there were no agricultural courses which could guarantee a supply.[10]

Liebig's research school at Giessen attracted an international clientèle which necessitated the director's being familiar with several languages, and as he spoke French and English, as well as his native German, Liebig fulfilled this condition. Boussingault, however, experienced difficulty with foreign tongues which is surprising since he came from a bilingual home. In Alsace he appealed to his German-speaking mother to keep house for him as he did not understand the locals, especially the servants. He admitted some difficulty in lecturing in Spanish at Bogotá although he had lived in a Spanish speaking country for nine months. A letter from J. H. Gilbert of Rothamsted dated April 27th, 1876, inviting him to England, contained an offer from Mrs Gilbert and Mrs Lawes to act as interpreters "if you do not bring an English-speaking friend".[11]

The publication of students' research papers is another condition for establishing a successful research school. Boussingault was ideally situated here as his own papers were readily accepted and for many years he shared, with others, the editorship of *Annales de chimie et de physique*. Thus, although he possessed some of the characteristics of a successful research school director he could never have succeeded in establishing and mainaining such an institution.

The Farmer

As a farmer at Bechelbronn he was undoubtedly aware of the need for an improved agriculture and pursued the scientific aspect of this goal for a major part of his career. Yet the evidence is that he was not a successful farmer nor was he particularly interested in the technical minutiae of farming operations. Noteworthy is the almost complete lack of practical farming information in his crop and animal husbandry papers. In this respect he can be compared with J. H. Gilbert of Rothamsted, but whereas Gilbert had the cooperation of J. B. Lawes, Boussingault relied on his brother-in-law, Frédéric Achille LeBel and a farm amnager, Eugène Oppermann, neither of whom showed any scientific inclination as did Lawes, who combined the skills of farmer and scientist.

The only critical assessment of the farm Bechelbronn, or at least Boussingault's portion of it, is contained in a work of Georges Ville published in 1868.[12] This was no personal attack on Boussingault as might be expected from previous encounters, but a survey of costs and production. Ville's survey not only included Bechelbronn but also the Grignon farm of the Institut agronomique and the farm school of an outstanding agronomist, Matthieu de Dombasle at Roville near Nancy. Ville's object was to demonstrate that these famous farms were never economically viable but would have been had the owners or managers used chemical fertilisers of which Ville had become an ardent advocate as a result of his trials at Vincennes on land given to him by Napoleon III. He insisted that to farm with only the manure from the farm stock was to farm without profit, and that the modest gain per annum at Bechelbronn did not allow for the farm manager's salary. Further, no sum was allocated to depreciation of buildings and equipment, the cost of manure had been assessed as the cost of production instead of the market price for such manure. Finally, the alleged farm profit, due to the sale of vegetable commodities, was produced with the aid of manure from stock occupying over one-half of the farm, land which might have been more profitably employed as arable land. Ville was not only an advocate of chemical fertilisers but also pleaded for a more critical accounting of farm income and expenditure, which in the case of Bechelbronn was almost certainly balanced by the oil receipts. Boussingault had believed, as he had stated in 1827 at Marmato that "we should demand from the soil the food necessary for the population", and with a growing population it is possible that he would have agreed with the employment of chemical fertilisers. Yet it is debatable whether Boussingault agreed with Ville's policy of excessive food production, or rather that of the soil yielding its rightful product, involving, as today, the opposing claims of agriculturist and conservationist.

Ville rightly regarded the farm as a productive unit and therefore could not make allowances for the use of farm labour and transport for the production of Pechelbronn oil, or that the farm was part of a greater complex. The evidence is not wholly conclusive but it does appear that Boussingault's engrossment in the scientific aspects of agriculture precluded an interest in the mechanics of farming which he, perhaps wisely, left to others.

The Scientist

As a scientist his work has been described in detail in the preceding chapters but his main achievements can be summarised as follows:

(a) The first ever complete analysis of the crops in a rotation.
(b) A legume crop increases the nitrogen content of the soil.
(c) The fat of a herbivore is derived from the carbohydrate in the ration (although the ultimate proof was demonstrated by Persoz).
(d) The optimum conditions of temperature and pressure by which oxygen may be obtained from the air on heating barium oxide and the subsequent decomposition of barium dioxide.
(e) The quantitative proof that the development of a plant is proportional to the amount of assimilatory nitrogen supplied to it, and that the greatest development occurs with the application of both nitrate and phosphate together. These conclusions were stated simultaneously by Georges Ville.
(f) The photosynthetic quotient is about unity.

Of these achievements, it is suggested that although the discovery concerning the photosynthetic quotient was perhaps his most accurate, brilliant and significant experiment, much of his career was devoted to the various stages of the nitrogen cycle in nature, the cycle which embraces the conversion of plant and animal debris and waste to nitrates in the soil and thence returned to plants and animals. His belief that the great ocean of air contributed to plant and animal nitrogen could not be verified but had to await the development of soil microbiology with its recognition of chemical changes due to bacterial action.

Boussingault was a prolific writer of scientific papers. South American subjects, with encouragement from Humboldt, occupied much of his professional life from 1825 to 1885, and it can be said that he wrote about 350 papers of which one-third were on the subject of South America, a quarter on crop husbandry, a seventh each on animal husbandry and chemistry, and the remaining eighth equally divided between metallurgy and botany. Although many of the South American papers could be placed in these other categories, the figure of one-third emphasises the powerful but unacknowledged influence of Humboldt. A perusal of his papers in chronological order reveals a distinct pattern of interests at various periods of his life:

1825–1835 South America;
1836–1848 Animal and crop husbandry;
1849–1851 Few papers – political activities intervened;
1852–1859 Nitrogen assimilation – plant nutrients;
1860–1883 Metallurgy of iron and steel – photosynthesis – soil nitrification.

In his scientific outlook and method he exhibited some professional weaknesses. Firstly, he was apt to ignore contemporary work and acquired "a somewhat harsh scepticism for the work of others which became habitual".[13] Secondly, with a few eclectic exceptions such as the work of Thaer and de Saussure, he appears not to have thought it worthwhile to review past work with the chance of a possible new interpretation leading to a fresh hypothesis. Thus, his vast and painstaking work on soil nitrification might have attained at least a partial conclusion had he displayed more sympathetic understanding of Pasteur's bacteriological work. Thirdly, a puzzling feature of Boussingault's rationale was his inability to construct conjectural models in spite of his tendency to seek analogies. A simple conjectural model is the explanation of some properties of molecules by comparing them with colliding billiard balls. A more relevant example, mentioned in a previous chapter, is the set of assumptions used by Hellriegel and Wilfarth in explaining the nitrification of the soil by legumes, by tendentiously positing a relationship between the root nodules of legumes and nitrogen fixing bacteria in the soil. In his student days at Saint-Etienne, Boussingault had hoped to confirm his analogy between the rôle of silicon and carbon in steel, and later to note the similarity between the syrupy excretion on the leaves of a lime tree at Liebfrauenberg and the manna described in Exodus 16,[14] yet this gift for analogising did not accompany the type of imagination required of an outstanding theoretician, an imagination which would have enabled Boussingault to construct and evolve hypotheses ready, if necessary, to be falsified by experiment.

Lastly, his 'open mind' attitude to his experiments appears strange; until an experiment was concluded and results obtained he would refrain from making any possible inferences, and even postpone any statement of results. He banished any preconceived ideas on the possible outcome of an experiment and once heatedly declared that when it is impossible to express a positive opinion we should know when to stop and say that we do not know, a sequel, perhaps, to his early lack of an advanced scientific education, the influence of which might have commended to him the policy of considering all known theories on the subject of the experiment undertaken, and all possible interpretations of the results. In another context, and discussing another discipline, it has been suggested that the 'open mind' attitude may be the hall-mark of the amateur.[15] With his polymathic inclinations Boussingault could not but help being an amateur in some branches of science and therefore approached his material without theoretical preconceptions.

Boussingault's Standing in the History of Science

An accepted scientific theory of agriculture was not possible until the discovery of nitrogen, an essential constituent of the protein of plant and animal tissue, so that agricultural theory as understood today could not have emerged until the latter part of the eighteenth century. A science at the pre-paradigm stage of development can be expected to exhibit various competing schools which are eventually resolved into a paradigm[16] with the elimination of those supporting implausible theories. A particular example was the humus theory, where it was believed that the sole source of plant carbon was derived from the humus in the top layer of the soil. The discovery of photosynthesis at the beginning of the nineteenth century appears not to have ousted the humus theory until the essential rôle of mineral salts was established. The pre-paradigm theories were resolved into a science of plant nutrition which took cognisance of the functions of humus, photosynthesis and mineral salts.

The competing schools of animal nutrition provide another example, the one mainly French and the other preponderantly German, where the discussion centred around the production of fat in a herbivore. The work of Boussingault and Persoz upheld the German viewpoint that fat was formed from carbohydrate foodstuffs and was not preformed in the plants eaten by herbivores. Generally, therefore, Boussingault's function in agricultural science was to help in creating a corpus of knowledge which in various ways is acceptable today; and to advance the subject from the stage of competing theories. To claim that he was the first agricultural chemist may be disputable but it can be said that his work resolved the formerly empirical nature of agriculture into a workable theory which pointed the way to future research. This was, indeed, an achievement.

In a brilliant age of French chemical theory, Boussingault was not a theoretician and it is, therefore, surprising that when August Kekulé, Carl Weltzein and Charles Wurtz combined to organise a meeting of chemists at Karlsruhe in Germany in 1860, with the object of resolving the chaotic state of formulae and atomic weights, the preliminary forty-five invitations included one to Boussingault. Further, he was elected chairman of the second day's meeting of the Karlsruhe Congress,[17] as it was called, where he made what might be regarded as an unnecessarily naive remark, "that it is not chemistry that grows old but chemists!"[18] On the third and last day of the Congress, which lasted from September 3rd to September 5th, 1860, under the chairmanship of Dumas, Boussingault insisted that the resolutions passed by the Congress need not be binding on individuals.[19] Yet at the Congress the professor of chemistry at the University of Genoa[20] had presented a lucid explanation of formulae with which most chemists concurred and which induced most of them to give up their arbitrary tables of atomic weights. But not Boussingault, and other French chemists, who continued to follow the practice of Berzelius (1779–1848) of Stockholm, who had been the leader of chemistry in Europe when they were young men. Berzelius, in his electrochemical or dualistic theory, had suggested that there was a significant connection between the electrical polarity and the chemical affinity of atoms. Attractive forces caused the atoms to combine to form compounds. His infleunce on the progress of chemistry in the first half of the ninetteenth century was immensely beneficial, bringing a sense of unity into the ever-growing number of chemical compounds. But his electrochemical dualistic theory oversimplified the subject and was found to be incapable of dealing with the structure of organic compounds. Liebig and Wöhler (1800–82) in 1832 showed that a complex group of atoms retained its identity as it passed through a series of compounds by the effect of chemical reactions. Such a group was called a 'radicle' and behaved like an atom in a compound. The radicle theory led to various theories of 'types' associated with a skilful galaxy of French chemists, Dumas, Laurent, Gerhardt and Wurtz. The history of nineteenth-century French chemistry emphasises the rôles of those who endeavoured to elucidate the structure of organic molecules, and not being one of these may account for Boussingault's comparative and unjustified neglect in the history of science which this biography has sought to remedy.

POSTSCRIPT

It was in May 1971 that my wife and I were motoring on the road to Saint-Etienne, having started that morning from Saint-Flour in the Cantal, that extraordinary town built on a rock pointing to the sky. Saint-Etienne was not unknown to us, having stayed there on a previous visit, and we knew it as a noisy industrial town which on this occasion we were anxious to avoid. I did, however, wish to see once again the school of mines which Boussingault attended as a youth. It now has a more worthy title as I was informed by a justly proud citizen from whom I enquired the way to the 'Ecole des Mines'. "You mean", he replied, holding himself stiffly upright, "l'Ecole Nationale Supérieure des Mines de Saint-Etienne?", for today it is an institution of university status.

The map tempted us to turn up a minor road west of Saint-Etienne and explore the neighbouring hamlets with such beautiful names evocative of the original unspoilt countryside, but alas, they turned out to be coal-mining villages to which travellers rarely ventured. So we continued along the tops of lovely hills bordering the young Loire until we decided to turn at St. Just-sur-Loire, fill up with petrol and head for Saint-Etienne which we were now unable to avoid.

It was at St. Just-sur-Loire that Mme Alice Crozet, the youngest of the Boussingault children, lived out her days as a widow on moving from Chambon-Feugerolles where her husband had managed a steelworks. She died in her ninetieth year in April 1928. I have before me the familiar printed notice of death which was sent to all relatives and close friends whose names are included in the notice. I have counted a hundred or so adults in addition to children, grandchildren, nephews and nieces whose names are not mentioned; also nine other families are informed so that altogether the possible number of mourners was considerable. Two familiar names are those of my good friends Mme Françoise Boussingault and M. Charles Boussingault, the great grand-children of Jean Baptiste Boussingault who were then teenagers. There is also Joseph Achille LeBel of stereoisomerism fame who was then eighty-one. It was he who, as a young man, went to the Froeschwiller battlefield to purchase horses to replace those commandeered by the Prussian forces in their advance through Bechelbronn in 1870. Another is Jean Louis

Boussingault (1883–1943), a son of Joseph, whose paintings have earned him almost as much space in the *Dictionnaire de Biographie Française* as his scientist grandfather. Yet it is the frequent recurrence of five names that interests the reader, the Holtzers, the Crozets, the Fourneyrons, the LeBels and, of course, the Boussingaults, representing a union of steel manufacture, civil engineering, chemistry and agriculture, witnesses of a gradual industrial revolution in France, a revolution still in progress in a land which, fortunately, maintains in most parts the appearance of a rural countryside.

APPENDIX 1

Syllabus of lectures given by J. B. Boussingault in agricultural chemistry at the Conservatoire des Arts et Métiers, 1862–1863.

DATE	SUBJECT MATTER
21st Nov.	The atmosphere.
23rd Nov.	The eudiometer, arrangement of apparatus.
25th Nov.	The atmosphere, physical composition. Distribution of heat on the earth's surface.
30th Nov.	Analysis of atmosphere by Bunsen's method.
5th Dec.	Winds, influence of seas on climate. Distribution of plants according to temperature of the vegetation zone. Cultivation subject to climate.
12th Dec.	Limits of vegetation in northern Europe, climate of the North Cape. Chemical constituents of the atmosphere. Composition of air.
14th Dec.	Analysis of air using phosphorus. Results obtained by the aid of this method.
19th Dec.	Chemical composition of the atmosphere (continued). Water in the atmosphere, its physical properties, freezing. Fields spoilt by the cold due to freezing of water which covers them. Formation of ice in rivers and in stagnant water.
21st Dec.	Analysis of atmospheric air by absorbants. General application of the method to a mixture of nitrogen, oxygen and a combustible gas. Estimation of carbonic acid in a gaseous mixture.
4th Jan.	Analysis of air by Brunner's method, and Dumas and Boussingault's method; substituting the balance for the eudiometer.
9th Jan.	Water vapour in the atmosphere. Chemical composition of water. Decomposition and synthesis of water.
11th Jan.	Eudiometric analysis. Detection and estimation of a combustible gas in a mixture containing mainly nitrogen and oxygen. Use of gas from the battery to measure combustion. Use of absorbant reagents in eudiometry. Application to testing and analysis of illuminating gas.

16th Jan. Properties of hydrogen. Surface water. Well water, river water. Air dissolved in water. Substances contained in water; water as a fertiliser in irrigation. Silt deposited by rivers. Volumes of water discharged by certain rivers.

18th Jan. Estimation of water vapour and carbonic acid in the air. Extraction and analysis of air in humus.

23rd Jan. Wet atmospheric conditions. Rain, mist, hail. Theory of dew formation, temperature fall at night, spring frosts, preventative measures. April moon.

30th Jan. Perpetual snows, glaciers, influence on irrigation. Dew. Freezing by the effect of nocturnal radiation. Effect of smoke.

1st Feb. Treatment of seed to prevent disease. A study of agents used. Method of detecting arsenic and for showing adulteration of substances used in seed treatment.

6th Feb. Carbon. Carbonic acid in the atmosphere. Origin of plant carbon.

8th Feb. Milk analysis. Methods and means of detecting the adulteration of milk.

20th Feb. Compounds of nitrogen and oxygen; nitric acid, nitrates, nitrification; sodium nitrate beds in Peru: nitrification of humus; application of theory of fallowing.

22nd Feb. Continuation of demonstrations of methods to detect the adulteration of milk.

27th Feb. Nitrates in river water and springs; application to irrigation. Nitrates in atmospheric water. Production of ammonium nitrate during storms. Ozone; its influence on nitrification.

1st Mar. General principles of sugar-cane and sugar-beet cultivation. Maple sugar, palm sugar. Methods of estimating sugar and glucose in sugar-producing plants.

6th Mar. Ozone in the atmosphere; its properties compared with those of oxygen. Nitrification by the influence of ozone. Organic materials in the atmosphere.

8th Mar. Continuation of methods of estimating sugar in plants. Estimating glucose in beetroot. Adulteration of sugar; methods of demonstrating it. The general cultivation of wheat and potatoes. Estimating starch in farinaceous plants.

13th Mar. Spontaneous generation. Organic germs in the atmosphere. Appert's method of preserving foods. Sulphur and phosphorus and their connection with plants and animals.

15th Mar. General facts about the cultivation of wheat and potatoes.

Estimation of starch in plants, for example, in wheat. Preparation of Schulze's reagent for detecting cellulose. Estimation of cellulose in potatoes.

20th Mar. Nutirition. General preservation of plant food and substances. Wheat taken as an example – starch, dextrin, gluten, albumin, casein. A comparison of plants with animals with regard to fibrin, albumin and casein.

22nd Mar. Estimation of cellulose and fats in plants. General considerations on the cultivation and production of oleaginous plants. Methods of demonstrating the adulteration of fuel oils. Analytical control of the products of an oil works.

27th Mar. On digestion. Gastric and pancreatic juices. Foods for body building and energy. Vegetable foods. Composition of forage taking hay as an example.

10th Apr. Physical and chemical composition of blood. Blood circulation. The phenomenon of respiration. Carbonic acid produced during respiration; man and horse. Volume of foul air; application to ventilation of houses and stables.

12th Apr. Coffee, its history, cultivation, adulteration and means of recognising it. Chicory; its manufacture.

17th Apr. Animal heat. Temperature of man and animals at the equator and polar regions. Transpiration, pulmonary and cutaneous. Urine of man and Herbivora.

19th Apr. Cultivation of the vine and the production of wine. Analysis of wine. Recommended methods of showing that wine has been adulterated. Sweetening wines.

24th Apr. Flesh and muscle. Cooking meat; boiling, roasting. Meat preservation. Fat considered as a food. Cheese. Butter.

APPENDIX 2

Syllabus of lectures given by J. B. Boussingault in agricultural and analytical chemistry at the Conservatoire des Arts et Métiers, 1873–1874.

DATE	SUBJECT MATTER
16th Nov.	Gas analysis. Application to agricultural research.
19th Nov.	Physical organisation of plants.
23rd Nov.	Analysis applied to agriculture. Continuation of gas analysis.
30th Nov.	Gas analysis continued.
3rd Dec.	Study of germination. Assimilation of carbon by plants.
7th Dec.	Gasometric analysis. Regnault's apparatus.
10th Dec.	Continuation of carbon assimilation. Plant respiration.
14th Dec.	Gravimetric analysis of a gas.
17th Dec.	Assimilation of nitrogen.
21st Dec.	Continuation of gravimetric analysis of gases, and elementary organic analysis.
24th Dec.	Assimilation of hydrogen and oxygen. Mineral substances in plants.
7th Jan.	The necessity for certain mineral substances in plants. The phenomenon of diffusion.
11th Jan.	Soil analysis; estimation of carbon and nitrogen.
14th Jan.	The state of mineral substances in the soil considered briefly, and their composition – for use in a theory of nutrition.
18th Jan.	Mineral analysis methods taken generally.
21st Jan.	Mineral nutrition by way of the roots.
25th Jan.	Qualitative analysis – the nature of potash, soda, baryta, strontia, lime.
28th Jan.	Study of soils. Agricultural geology.
1st Feb.	Qualitative analysis – the nature of magnesia, alumina, iron, manganese, common mineral acids.
4th Feb.	Relationship between geology and agriculture. A study of factory grounds and alluvium.

8th Feb.	Nature of acids (continued). Estimation of water. A thermal mixer ('thermo-agitateur') for gases.*
11th Feb.	Study of soils. Physical properties of clay and mould.
22nd Feb.	Qualitative analysis; water content; estimation of potash and soda.
25th Feb.	Different methods of physical analysis of arable lands. Considerable errors in estimating clay by the old method. The new method.
1st Mar.	Alkalimetry. Estimation of ammonia.
4th Mar.	Physical properties of arable lands.
7th Mar.	Demonstration continued of alkalimetric methods and the estimation of ammonia.
11th Mar.	Continuation of the physical properties of soil. External influences. Classification.
14th Mar.	Estimation of baryta, lime, magnesia and alumina. Method of M. le Cte de Deville.
18th Mar.	Absorbent properties of soilds.
22nd Mar.	Analytical methods of M. Ste. Claire Deville.
25th Mar.	A study of the decomposition of substances in the soil.
1st Apr.	The rôle of oxygen in the soil. The atmosphere confined in the soil.
8th Apr.	Study of the nitrification of arable land.
12th Apr.	Estimation of silicic acid, free or as silicates. Carbonic acid.
15th Apr.	Nitrification studies in arable land.
19th Apr.	Estimation of acids: hydrochloric, sulphuric, nitric.
22nd Apr.	Recapitulation of the principal facts set forth so far.

* The 'thermo-agitateur' mentioned by Boussingault perhaps refers to convection currents in the atmosphere resulting in rainfall.

ABBREVIATIONS

A.C.	*Annales de chimie et de physique*
A.C.A.M.	*Annales du Conservatoire des Arts et Métiers*
A.C.P.	*Agronomie, chimie agricole et physiologie*
A.M.	*Annales des Mines*
A.S.B.	Académie des Sciences, Boussingault dossier
A.S.H.	Académie des Sciences, Humboldt dossier
A.S.N.	*Annales des sciences naturelles* (followed by Bot. or Zool.)
B.G.F.	*Bulletin de la société géologique de France*
B.S.A.	*Bulletin des séances de la société d'agriculture de France*
C.R.	*Comptes rendus hebdomadaire des séances de l'Académie des Sciences*
J.A.P.	*Journal d'agriculture pratique* (La maison rustique du 19e siècle)
J.P.	*Journal de pharmacie et de chimie*
Jl R. Agric. Soc.	*Journal of the Royal Agricultural Society*
M.A.C.	*Archives de la commission scientifique du Mexique*
Mém.	*Mémoires de J. B. Boussingault*
M.S.A.	*Mémoires de la société national d'agriculture*
Phil. Trans. R. Soc.	*Philosophical Transactions of the Royal Society*
PVCP	*Procès-verbaux du Conseil de Perfectionnement du Conservatoire des Arts et Métiers*
Rep. Br. Ass. Advmt Sci.	*Reports of the British Association for the Advancement of Science*

Unless otherwise stated, the journals and volumes quoted were published in London or Paris.

NOTES

Introduction

1. Cardus, R.: 1945, *Ten Composers*, p. 108.
2. Lacroix, A.: 1926, *Notice historique sur Jean Baptiste Boussingault*, reprinted in *Figures de Savants*, Vol. 2 (1937).
3. Dehérein, P. P.: 1887, 'L'oeuvre agricole de M. Boussingault', *Revue scient. 2*, pp. 33–48.
4. Salet, G.: 1888, 'J. B. Boussingault', *Revue scient. 4*, p. 373.
5. Schloesing, T. *et al.*: 1895, *Inauguration du monument de Boussingault au Conservatoire des Arts et Métiers à Paris le Dimanche 7 juillet 1895.*
6. Leouzon, L.: 1905, *Agronomes et éleveurs*, pp. 317–331.
7. Lenglen, M.: 1937, *J. B. Boussingault à la société centrale d'agriculture* (Beauvais).
8. Lenglen, M.: 1940, *Boussingault à Bechelbronn et au Liebfrauenberg* (Alais).
9. LeRoux, D.: 1936, 'J. B. Boussingault créateur de la chimie agricole, sa vie et son oeuvre au Conservatoire Nationale des Arts et Métiers', *Bull. de l'assoc. française pour l'étude du sol 2*, pp. 246–254.
10. Prijanischnikow, D. N.: 1936, 'Cent ans de chimie agricole', a translation from *Izvestia* of May 30th, 1936, in *Bull. de l'assoc. française pour l'étude du sol 2*, pp. 246–254.
11. Lavollay, J.: 1961, 'Boussingault, les Schloesing et les sciences de bas de l'agronomie', *Comptes rendus de l'Acad. d'Agriculture de France 47*, pp. 233–239.
12. Cowgill, G. R.: 1964, 'Jean Baptiste Boussingault – a biographical sketch', *Jl nutrition 84*, pp. 1–9 (Philadelphia).
13. McCollum, E. V.: 1957, *A History of Nutrition* (Boston).
14. Aulie, R. P.: 1968, *Boussingault and the nitrogen cycle*, Yale Ph.D. thesis.
15. Aulie, R. P.: 1970, 'Boussingault and the nitrogen cycle', *Proc. Amer. Phil. Soc. 114*, No. 6, pp. 435–479.
16. Aulie, R. P.: 1974, 'The Mineral Theory', *Agricultural History 48*, No. 3, pp. 369–382 (Univ. of California Press).
17 Rossiter, M. W. 1975, *The Emergence of Agriculture Science, Justus Liebig and the Americans* (Yale Univ. Press), Chaps. 2 and 3.

Chapter 1

1. Much of the early biographical material is taken from *Mémoires de J. B. Boussingault*, 5 vols. (1892–1903), Volume 1.
2. The 17th-century spelling of the name was Bouçingo. It was written as Boisimcolt in the marriage register at Wetzlar, an error no doubt as his father, Charles Nicholas, signed letters as Boussingault, a spelling which has been observed since. Yet it was

written as Boussaingault in the royal ordinance of 1839 confirming Boussingault's election to the Académie des Sciences.

3. *Mém. 1*, 28.
4. *Ibid.*, 27.
5. *Ibid.*, 2.
6. *Ibid.*
7. *Ibid.*, 33.
8. The Committee of Public Safety under Robespierre had organised a well disciplined army of 850 000 men to repel invasions from neighbouring countries. Under generals who risked execution as the price of failure, enemy armies were routed including that of the Austrian Netherlands which was pushed by Jourdan beyond the Main so that the French forces occupied the Palatinate. Jourdan's southern flank was held by the French army of the Rhine and Moselle under Pichgru who treacherously arranged with the opposing Austrian general that his army would march on Paris and restore the Bourbons. Thus Jourdan was compelled to retreat.
9. *Mém. 1*, 33.
10. The Münch family history is well documented in the Historical Archives of the city of Wetzlar.
11. *Mém. 1*, 33.
12. *Ibid.*, 26.
13. *Ibid.*, 2.
14. Hillairet, J.: 1952, *Evocation de Vieux Paris 1*, 535.
15. Lecouzon, L.: 1905, *Agronomes et éleveurs*, 318.
16. *Mém. 1*, 27.
17. *Ibid.*, 3. J. César LeGallois (1770–1814).
18. *Ibid.*, 10, 16.
19. *Ibid.*, 39.
20. *Ibid.*, 39–40.
21. Crosland, M. P.: 1967, *The Society of Arcueil*, 53–54.
22. *Mém. 1*, 41.
23. *Ibid.*, 42–43.
24. *Ibid.*, 74–75.
25. *Ibid.*, 44–45.
26. Crosland, M. P., 1978, *Gay-Lussac Scientist and Bourgeois*, 146–147.
27. *Mém 1*, 45.
28. *Ibid.*, 50.
29. *Ibid.*, 49.
30. *Ibid.*, 25.
31. *Ibid.*, 85.
32. *Ibid.*, 94.
33. *Ibid.*, 102.
34. *Ibid.*, 250, Boussingault to father 28th May, 1821.
35. Institut de France, Ms 2420 B361, Boussingault to Auguste Mabru Director of Mines, Pechelbronn 29th March, 1828.
36. Collection Mme F. Boussingault, Les Montils, Loir-et-Cher. J. H. Gilbert to Boussingault, 27 April, 1876.
37. Ecole Nationale Supérieure des Mines de Saint-Etienne, *Annuaire* 1965, 19.

38. *Mém. 1*, 103.
39. *Ibid.*, 192–196. Boussingault to parents, 7 Dec., 1818 and 11 Dec., 1818.
40. *Mém. 2*, 238–239.
41. *Le Centenaire de l'Ecole Nationale des Mines de Saint-Etienne* 1921, 13–14, 39 (Saint-Etienne).
42. *Mém. 1*, 198–222 *passim*.
43. *Op. cit.*, Note 37 *passim*.
44. *Mém. 1*, 195–197. Boussingault to parents, 11 Dec., 1818; Boussingault to father, 16 Dec., 1818.
45. *Ibid.*, 205. Within the family, Boussingault was known as 'Lolo', the small child's rendering of 'lait-lait' although there is no record that he was unusually fond of milk.
46. *Ibid.*, 199. Boussingault to parents, 22 Dec., 1818.
47. *Ibid.*, 214–215.
48. Thibaud, M.: 'Des procédés suivis à Chessy pour la préparation mécanique des minerais', *A. M.*, 1827, *3*, 195–207.
49. *Mém. I*, 204. Boussingault to parents, 21 April, 1819.
50. 'Note sur le sel ammoniaque que produit une mine de houille incendiée par les Elèves mineurs de Saint-Etienne', *A. C.*, 1822, *21*, 158–159.
51. *Mém. 1*, 133–138.
52. *Ibid.*, 217–220. Boussingault to father, 26 Feb., 1820 and 24 April, 1820.
53. *Ibid.*, 215. Boussingault to father 24 Dec., 1819.
54. *Op. cit.*, Note 52, letter of 26 Feb.
55. *Ibid.*, 224–226. Boussingault to father, 18 May, 1820.
56. 'Note sur la combinaison du silicium avec le platine, et sur sa présence dans l'acier', *A. C.*, 1821, *16*, 5–16.
57. *Journal des Mines*, 1798, *8*, 707.
58. *Mém. 1*, 265. Beaunier to Boussingault, 30 Aug., 1821.
59. Stodart, J. and Faraday, M.: 'On the alloys of steel', *Phil. Trans.* 1822, *112*, 253.
60. *Mém. 2*, 256. Boussingault to father, 22 Mar., 1820.
61. *Mém. 1*, 205. Letter to parents, 21 Apr., 1819.
62. *Mém. 2*, 241. Father to Boussingault, 5th April, 1819.
63. *Mém. 1*, 220. Boussingault to father, 24 Apr., 1820 and 215 of 24th Dec., 1819.
64. *Mém. 2*, 249. Father to Boussingault, 5 Dec., 1819.
65. *Ibid.*, 278. Mme Vaudet to Boussingault, 22nd July, 1821.
66. *Mém. 1*, 9.
67. *Ibid.*, 208. Boussingault to father, 8 May, 1819.
68. *Mém. 2*, 259. M. Vaudet to Boussingault, 11 Apr., 1820.
69. *Ibid.*, 270. Mother to Boussingault, 24 June, 1820.
70. *Mém. 1*, 265. Beaunier to Boussingault, 30 Aug., 1821 enclosing his certificate from the mining school at Saint-Etienne.

Chapter 2

1. *Mém. 1*, 221–222, 228–229. Boussingault to father, 20 Apr., 1820 and 3rd July, 1820.

2. *Ibid.*, 151–153.
3. An authoritative account of his life and achievements is found in Crosland, M. P.: 1978, *Gay–Lussac Scientist and Bourgeois* (C.U.P.).
4. *Mém. 1*, 234–235. Boussingault to father, 18 Sept., 1820.
5. *Ibid.*, 236–237. Boussingault to father, 20 Sept., 1820.
6. *Ibid.*, 239–242. Boussingault to father, 9 Feb., 1821.
7. *Mém. 2*, 286–287. Mme Vaudet to Boussingault 22 Jan., 1822.
8. See Note 5.
9. *Mém. 1*, 249. Boussingault to father, 28 May, 1821.
10. *Ibid.*
11. *Ibid.*, 252. Father to Boussingault, 8 June, 1821.
12. *Ibid.*, 159.
13. Branthôme, Abbé: 1826–27, *Chimie théorique et pratique*, 236 pp. MS 3177 (557) 19 at Bib. nat. et univ. Strasbourg.
14. *Ibid.*, 25.
15. See Appendix 1.
16. Syndicat d'Initiative, Merckwiller-Penchelbronn, *La source thermale de Pechelbronn* (mimeographed).
17. Chambrier, Paul de: 1919, *Historique de Pechelbronn 1498–1918* (Neuchâtel) 41: 1921, *Exploitation du pétrole par puits et galeries*, 9.
18. *Mém. 1*, 243–244, 245–247 and 257–259. Letters from Thibaud 17 May, 27 May and 26 June, 1821.
19. *Ibid.*, 267–269. Gueynivaud to Boussingault, 1 Feb., 1822. Gueynivaud was a lecturer at the Ecole des Mines, Saint-Etienne.
20. *Ibid.*, 168, 270. Humboldt to Boussingault, July 1822; 271–272. Humboldt to Boussingault, 5 Aug., 1822.
21. *Ibid.*, 275. Humboldt to Boussingault, 14 Aug., 1822.
22. *Ibid.*, 177.
23. *Ibid.*, 178–181.
24. *Ibid.*, 182–183.
25. *Ibid.*, 189.
26. *Ibid.*, 191 and 277.
27. Karady, V. 'Educational qualifications and university careers in science in nineteenth century France', Fox, R. and G. Weisz (eds.): 1980, *The Organization of Science and Technology in France 1808–1914*, 97.
28. *Mém. 1*, 168.

Chapter 3

1. *Mém. 2*, 1–5.
2. Combes, M.: 1929, *Roulin et ses amis*, 42–55.
3. Cochrane, C. S.: 1825, *Journal of a Residence and Travels in Colombia during the years 1823–1824*, 2 vols.; Hamilton, J. F.: 1827, *Travels through the Interior Provinces of Colombia*, 2 vols., esp. *1*, 182.
4. Humphries, R. A.: 'The Emancipation of Latin America', in *The New Cambridge Modern History*, Cambridge, 1965, *9*, Chap. 23, 612–628, to which reference has

been made for the salient facts in the liberation of Venezuela and New Granada.

5. Boussingault, J. B. and Rivero, M. E. de: *A. C.*, 1824, *25*, 427–429.
6. *Mém. 3*, 273. Boussingault to mother, 18 June, 1823.
7. *Mém. 2*, 156.
8. Singer, C. and Underwood, E. Ashworth: 1928, *A Short History of Medicine* (Oxford), 2nd ed. 1962, 520–530.
9. Parra, H.: 'Simple Goitre in Colombia', *Amer. J. publ. Hlth 38*, 820.
10. *A. C.*, 1833, *54*, 176–177.
11. *Ibid.*, 176.
12. *A. C.*, 1831, *48*, 176.
13. *A. C.*, 1825, *30*, 91.
14. Note 9, 820.
15. Note 13, 92; *Mém. 4*, 116.
16. *Mém. 2*, 185–186.
17. *A. C.*, 1825, *29*, 110–111.
18. $Na_2CO_3 . CaCO_3 . 5H_2O$.
19. *Mém. 2*, 138–139.
20. (a) *Ibid.*, 196–199; (b) Boussingault and Rivero, *A. C.*, 1824, *25*, 438–443.
21. *Mém. 2*, 214–234.; *Mém. 4*, 351–352. Boussingault to Uncle Louis, 9 Dec., 1824.
22. Inst. de France, MS 2420 B 361. Boussingault to Auguste Mabru, 29 Mar., 1828.
23. *Mém. 3*, 119–127.
24. *Ibid.*, 245.
25. *Mém. 1*, 149.
26. *Mém. 3*, 245.
27. *A. S. H.* Humboldt to Boussingalt, 10 Oct., 1824.
28. *Mém. 4*, 18–25.
29. *A. C.*, 1834, *57*, 146–182.
30. Liebig, J. von: 1840, *Organic Chemistry in its Applications to Agriculture and Physiology* (trans. L. Playfair), 375–376.
31. *Mém. 4*, 5–6.
32. McIntyre, A. R.: 1947, *Curare, Its History, Nature and Chemical Use* (Chicago), 5.
33. Humboldt, A. von and Bonpland, A.: *Personal Narrative of Travels to the Equinoxial Regions* (trans. N. E. Williams), 1821, 497 cited by McIntyre, Note 32.
34. *Mém. 2*, 69.
35. Boussingault et Roulin: *A. C.*, 1828, *39*, 24–37; Pelletier and Petros, *A. C.*, 1829, *40*, 213–219, made a further examination of curare but admitted that they had nothing to add to the research results of Boussingault and Roulin.
36. Roulin, F.: 1865, *Histoire naturelle et souvenirs de voyage.*
37. *Mém. 1*, 167.
38. See Note 35, Chap. 1.
39. *Mém. 3*, 24.
40. *Mém. 4*, 42.
41. *Mém. 5*, 80.
42. *Mém. 3*, 24.
43. *Ibid.*, 46.
44. *Ibid.*, 220–221.

45. *Mém. 4*, 81–83.
46. *Ibid.*, 192.
47. *Ibid.*, 113.
48. *Mém. 3*, 276. Boussingault to mother, 18 Jan., 1823.
49. (a) MacDonald, D.: 1960, *A History of Platinum*, 14–19; MacDonald, D. and L. B. Hunt: 1982, *A History of Platinum and Its Allied Metals*, 15; (b) *Mém. 5*, 7.
50. *Mém. 5*, 4.
51. *Mém. 4*, 260–261.
52. *Mém. 4*, 245–248.
53. *A. C.*, 1826, *32*, 204–212.
54. *Mém. 2*, 54–56. Dr J. A. Chaldecott has kindly drawn my attention to a letter, *Mém. 3*, 282–283, in which Vaudet suggests that Boussingault might export platinum to Europe. There is no further reference to this suggestion.
55. (a) Baker, J. N. L.: 1931, *A History of Geographical Discovery and Exploration*, repr. 1948, 404; (b) Parias, L. H.: 1956, *Histoire universelle des explorations 4*, 145.

Chapter 4

1. *Mém. 3*, 175, 178 and 222.
2. *Mém. 5*, 24–30.
3. *Mém. 4*, 148.
4. *Mém. 5*, 58–59.
5. *A. C. 58*, 150–180.
6. *Mém. 5*, 154, 156.
7. *Ibid.*, 73.
8. Gay-Lussac, J. L.: 1823, *A. C. 22*, 415–429, esp. 421.
9. Lacroix, A.: 1932, *Figures de Savants 2*, 296.
10. Smeaton, W. A.: 1965, *Ambix 13*, No. 1, 84–91.
11. *Mém. 5*, 69.
12. *Ibid.*, 192–217, 243.
13. *Ibid.*, 250–253.
14. See Note 5.
15. *A. S. H.* Humboldt to Boussingault, 24 Oct., 1834.
16. Whymper, E.: 1892, *Travels amongst the great Andes of the Equator*, 23–24, 76.
17. *C. R.*, 1860, *51*, 844–855. There is some slight evidence that Boussingault visited the Galapagos islands (Lenglen, M.: 1937, *J. B. Boussingault à la Société Centrale d'Agriculture* (Beauvais), 39.).
18. *Mém. 5*, 332–333 and Figure 10.
19. *C. R.*, 1845, *21*, 1002, 1149–1153.
20. Humboldt's letters to Boussingault, covering the period 1822–1858, in the Humboldt dossier of the *Académie des Sciences*, were collected by a grandson, Edmund Boussingault.
21. *A. S. H.* Humboldt to Boussingault, Part I, 77, 81, of 18 Dec., 1832; Part 2, 15, of 12 Sept., 1834.
22. *Inauguration du Buste de J. B. Boussingulat à l'Hôtel de la Société amicale des*

anciens éleves de l'Ecole nationale des Mines de Saint-Etienne le 14 juin 1913 (Saint-Etienne) mentions Mme Crozet (Alice) as his most valuable collaborator in writing the *Mémoires*, but her husband, Jean Crozet, assisted as revealed by a letter from Boussingault of 22 Mar., 1881 (collection Mme F. Boussingault, Les Montils). A secondary source mentions that Boussingault's elder daughter, Berthe Holzer, edited the *Mémoires* (Chardon, Charlos E.: *Boussingault* (Ciudad Trujillo, 1953), 64).

23. *A. C.*, 1832, *51*, 337–356; also Schnabel, C.: 1898, *Handbook on Metallurgy* (trans. H. Louis), 2 vols., *1*, 625, 637–638; and Roscoe, H. E. and Schorlemmer, C.: 1913, *A Treatise on Chemistry*, 5th ed., *2*, 450–452.
24. *A. C.*, 1833, *53*, 225–247; *C. R.*, 1835, *1*, 385–387.
25. *A. C.*, 1827, *34*, 408–419; *C. R.*, 1831, *4* 950–956.

Chapter 5

1. Inst. de France MS 2420 B364. Boussingault to Voltz 12 July, 1834.
2. *A. S. H.* Pt. 1, 85. Humboldt to Boussingault, 18 Dec., 1832.
3. *Mém. 5*, 368–369. Vaudet to Boussingault, 27 Mar., 1826.
4. *A. S. H.* Pt. 2, 2. Humboldt to Boussingault, 2 Feb, 1833.
5. *Ibid.*, Pt. 1, 80. Humboldt to Boussingault, 18 Dec., 1832.
6. *C. R.*, 1838, *6*, 149. Votes were cast for Audouin (38), Gasparin (17), Soulange-Bodin (1) but 10 others including Boussingault received no vote.
7. *A. S. H.* Pt. 2, 51. Humboldt to Boussingault, undated, probably 1838.
8. *C. R.*, 1839 8, 131–132; voting was Boussingault (40), Payen (11), Gasparin (1), two others received no vote.
9. Humboldt, A. von: 1836–39, *Examen critique . . . seizième siècles*, xx–xxi.
10. *A. S. H.* Pt. 2, 7. Humboldt to Boussingault, 16 May, 1834.
11. *Ibid.*, 15, Humboldt to Boussingault, 12 Sept., 1834.
12. Registers of the commune of Lampertsloch, kept at the Catholic school.
13. *A. S. H.* Pt. 2, 24. Humboldt to Boussingault, 4 Dec., 1834.
14. *A. S. B.* Mme Crozet-Boussingault to M. Pierre Gauja, July 1926.
15. Dutacq, F.: 1939–1940, *Annales de l'Université*, 59.
16. *Livret de l'Etudiant*, 1967 (Univ. de Lyon), 7.
17. Dutacq, Note 15, 62.
18. Cobban, A.: 1961, *A History of Modern France*, 2nd ed. 1965, *2*, 93.
19. Bard, J.: 1842, *Le Palais St. Pierre ou des Arts* (Lyon), 9; Hervier, M.: 1922, *Les Palais des Arts* (Lyon), 54; Ehrhard, A.: 1919, *l'université de Lyon* (Lyon), 135.
20. *Le Courrier de Lyon* (appearing thrice weekly), 2 Sept., 1834, 1.
21. *Revue d'Histoire de Lyon*, 1905 (Lyon), 4.
22. Arch. dep. Rhône, N, 1832, Rapport d'introduction du Prêfet (cited in *Livret de l'Etudiant*, Note 16, 7).
23. *Le Courrier de Lyon*, 1 Jan., 1835, 2.
24. *Le Courrier de Lyon*, 31 July, 1834.
25. *A. S. B.* Minister of Public Instruction to Rector, Académie de Lyon, 4 Nov., 1834.
26. *Le Courrier de Lyon*, 30 Jan., 1835; *Le Reparateur* (an opposition journal), 30

Jan., 1835; *l'Athénée*, 1835, 40–41. This was a weekly literary and scientific magazine which appeared each Sunday from January to May 1835 when it ceased publication.

27. Artz, F. D.: 1966, *The Development of Technical Education in France 1500–1850* (MIT, Camb., Mass.), 212–214.
28. *A. S. B.* Boussingault to Soulacroix, 28 Feb., 1835.
29. *A. S. B.* Prof. Offert of Lyon to Alfred Lacroix, *Secrétaire Perpétuel de l'Académie des Sciences*, 28 Jan., 1925.
30. *A. S. B.* Boussingault to Rector (Soulacroix), 29 Oct., 1836. The correspondence between Boussingault, the Rector of the *Académie de Lyon*, and the Minister of Public Instruction was found after a prolonged search in 1924 by the professor of mineralogy, Albert Offret, of the University of Lyon at the request of Alfred Lacroix, who was preparing a lecture, *Notice historique sur Jean Baptiste Boussingault* for the annual public meeting of the *Académie des Sciences* on 13 Dec., 1926. This correspondence has been lost but fortunately Offret made summaries of most of the letters and copied several important ones and all these are in the Boussingault dossier at the *Académie des Sciences*.
31. *A. S. B.* Rector to Minister, 18 Mar., 1834.
32. *Ibid.*, Minister to Rector, 18 June, 1835.
33. *Ibid.*, Rector to Minister, 20 Jan., 1835.
34. *Ibid.*, Rector to Minister, 10 Mar., 1835.
35. *Ibid.*, Minister to Rector, April 1835.
36. *Le Courrier de Lyon*, 8 Nov., 1835, 2.
37. Pouzet, Ph.: 1937, *Le Passé du Lycée Ampère* (Lyon), 32.
38. *A. S. B.* Minister to Rector, 20 Nov., 1835.
39. *Ibid.*, Rector to Minister, 29 Dec., 1835.
40. *Le Courrier de Lyon*, 30 Dec., 1835, 2.
41. *A. S. B.* Minister to Rector, 25 Dec., 1835.
42. Crosland, M.: 1967, *The Society of Arcueil*, 119.
43. *A. S. B.* Boussingault to Rector, 21 Jan., 1836.
44. *Ibid.*, Boussingault to Rector, 26 Jan., 1836.
45. *Ibid.*, Boussingault to Rector, 1 Mar., 1836.
46. *Ibid.*, Rector to Minister, 21 Feb., 1836.
47. *Ibid.*
48. *Ibid.*, Ministerial order of 31 Mar., 1836 to Rector, appointing Regnault.
49. *Ibid.*, Ministerial order of 24 Nov., 1836 to Rector appointing Bineau.
50. *Le Courrier de Lyon*, 3 June, 1836, 2.
51. *A. S. B.* Note 45.
52. *A. S. B.* Rector to Minister, 10 April, 1836.
53. *Ibid.*, see Note 45.
54. *A. S. H.* Pt. 2, 42. Humboldt to Boussingault, 14 May, 1837.
55. Johnson, D.: 1963, *Guizot*, 154.
56. *C. R.*, 1835, *1*, 87–89; *A. C.*, 1834, *57*, 161.
57. *Le Courrier de Lyon*, 18 July, 1836; *Revue de Lyonnais*, 1838, *7*, 497–498.
58. Boussingault, J. B.: 1844, *Economic rurale*, 2 vols., *2*, 254.
59. *A. C.*, 1836, *63*, 337–358.
60. Sachs, J. von: *History of Botany (1530–1860)* (trans. H. E. F. Garney), Oxford,

2nd imp. 1906, 560–562. Multitudinous recordings of temperature and growth were made and this activity became a branch of Phenology, a term first used in 1862 for the recording and study of the dates of recurrent natural events (*Encycl. Brit.*, 1971, *17*, 806).

61. *Registre des Procés-Verbaux des Actes et déliberations de la Faculté des Sciences de l'Académie de Paris*, 26 Oct., 1835.
62. A copy of this letter was kindly supplied by the Conservateur aux Archives Nationales.
63. *C. R.*, 1841, *12*, 1005–1025; *A. C.*, 1841, *3*, 257–318.
64. *C. R.*, 1844, *18*, 473–476.
65. *C. R.*, 1841, *13* 366; *A. C.*, 1844, *10*, 456–469.
66. *A. C.*, 1839, *71* 128–136.

Chapter 6

1. *A. C.*, 1827, *34*, 253–268.
2. *Mém. 4*, 196.
3. *Agronomie, chimie agricole et physiologie*, 1861, *2*, 372–377.
4. *C. R.*, 1860, *51*, 844–855.
5. Wackermann, C.: 1951, *Le champ pétrolifière de Pechelbronn* (Travaux du laboratoire de Géographie de l'Université de Strasbourg).
6. *Etat nominatif des habitants* – commune de Lampertsloch, 1836, Archives du Bas-Rhin, Strasbourg, No. 7M 489.
7. *A. C.*, 1845, *14*, 433.
8. *Georgics 1*, 73–75, Loeb Classical Library (William Heinemann Ltd.)
9. Duhamel du Monceau (usually referred to as 'Duhamel'), 1750–1756; *Traité de la culture des terres, suivant les principes de M. Tull Anglais*, 6 vols.
10. Bourde, André J.: 1953, *The Influence of England on the French Agronomes, 1753–1789* (Cambridge), Chap 6.
11. Thaer, A. D.: 1846, *The Principles of Agriculture* (New York), 2 vols., *I*, 132–143 (trans. W. Shaw and C. Johnson from *Grundsatze der Rationellen Landwirthschaft*, 1809–1812 (Berlin), 4 vols.).
12. *Bulletin de l'Association française pour l'étude du Sol*, 1936, *2*, 247–248.
13. (a) Boussingault, J. B.: 1845, *Rural Economy* (trans. G. Law), 473–474, from *Economie rurale*, 1841–1842; (b) *C. R.*, 1838, 7, 1149–1155.
14. *A. C.*, 1844, *2*, 308–321.
15. *C. R.*, 1839, *8*, 54–56.
16. Liebig, J. von; 1840, *Organic Chemistry in Its Applications to Agriculture and Physiology* (trans. L. Playfair), 99.
17. *Ibid.*, 176. Liebig conceded that Boussingault's analysis "merited the greatest confidence" (*Ibid.*, 177).
18. *Ibid.*, 3 rd ed. 1843, 203–204.
19. *Economie rurale*, 2nd ed., *I*, 82.
20. (a) Hall, A. D.: 1905, *The Book of the Rothamsted Experiments*, 5; (b) Lawes, J. B.: 1847, *Jl R. agric. Soc. 8*, 495–565.
21. Hall, see Note 20 (a).

22. Liebig, see Note 16, 3rd ed., 211.
23. Lawes, see Note 20 (b).
24. *C. R.*, 1841, *13*, 323–332, 327.
25. Lenglen, M.: 1937, *J. B. Boussingault à la Société Centrale d'Agriculture* (Beauvais), 93.
26. Registers of the commune of Lampertsloch, kept at the Catholic school.
27. *A. S. B.* Boussingault to Dumas, 1 Aug., 1842.
28. These memoirs were eventually included in *Mémoires de chimie agricole et de physiologie*, 1854.
29. *A. S. B.* Boussingault to Dumas, 26 May, 1844.
30. Collection Mme F. Boussingault, Les Montils, Liebig to Boussingault, May 1844.
31. Boussingault, J. B.: *Economie rurale 2*, 14.
32. Voelcker, A.: 1856, *Jl. R. agric. Soc. 17*, 191.
33. Lenglen, see Note 25, 101.
34. *A. C.*, 1841, *I*, 216.
35. LeClerc-Thouin, O.: 1835, *Notions théoriques et pratiques sur les assolements*, 259.
36. Lenglen, see Note 25, 25–26.
37. Candolle, A. P.: 1832, *Physiologie végétale*, 248–251.
38. It was suggested by J. Macaire-Prinsep that the plant roots excreted toxic materials; this theory was mentioned by Liebig, see Note 16, 163.

Chapter 7

1. *A. C.*, 1816, *3*, 68–77.
2. Prout, W.: *Phil. Mag.*, 1827, *2*, 144; 1828, *3*, 33, 107, cited by McCollum, E. V.: 1957, *A History of Nutrition* (Boston), 88, 98.
3. Thaer, A. D.: 1844, *The Principles of Agriculture* (trans. W. Shaw and C. Johnson), 2 vols., *1*, 165–172.
4. (a) *A. C.*, 1836, *63*, 225–244; (b) *A. C.*, 1838, *67*, 408–421.
5. *A. C.*, 1789, *3*, 252, cited by Partington, J. R.: 1962, *A History of Chemistry, 3*, 546.
6. *A. C.*, 1837, *65*, 301–320.
7. *A. C.*, 1839, *71*, 65–79.
8. *Ibid.*, 113–117.
9. *Ibid.*, 128–136.
10. Lawes, J. B. and Gilbert, J. H.: 1852, *Rep. Br. Ass. Advmt Sci.*, 323–353, esp. 352; *Jl R. agric. Soc.*, 1860, *21*, 488.
11. (a) *A. C.*, 1943, *8*, 63–66; (b) *C. R.*, 1843, *16*, 347.
12. See Note 11 (b), 351.
13. *C. R.*, 1849, *28* 249–253.
14. *C. R.*, 1843, *16*, 559–560.
15. *Ibid.*, 556.
16. *Ibid.*, 557.
17. Huber, F.: 1792, *Nouvelles observations sur les abeilles* (Geneva), *2*, Chap. 1, cited by Dumas and Milne Edwards, *A. C.*, 1845, *14*, 400.

18. Gundelach, F. W.: 1842, *Die naturgeschichte der Honigbienin* (Cassel), cited by Dumas and Milne Edwards (Note 17), 401.
19. See Note 14, 664.
20. *A. C.*, 1845, *14*, 419 482, 430.
21. *A. S. B.* Boussingault to Dumas, 13 Jan., 1845.
22. *Ibid.*
23. *C. R.*, 1845, *20*, 1726.
24. *C. R.*, 1844, *18*, 245–254.
25. *A. S. B.* See Note 21.
26. *Ibid.*
27. See Note 14, 664.
28. *C. R.*, 1845, *20*, 1726.
29. *Jl. R. agric. Soc.*, 1860, *21*, 488.
30. *C. R.*, 1838, *7*, 1020.
31. Boussingault, J. B.: 1845, *Rural Economy* (trans. G. Law), 633.
32. Liebig, J. von: 1843, *Phil. Mag. 23*, 22.
33. Lenglen, M.: 1940, *Boussingault à Bechelbronn et au Liebfrauenberg* (Alençon), 14.
34. *Ibid.*, 11. "Boussingault à Bechelbronn – au premier chimiste agricole ses continuateurs adressent un chaleureux vivat".
35. Bib. de l'Institut de France MS 15 (2043) 48. Boussingault to Arago, 13 July, 1847. This was the first letter from Liebfrauenberg.

Chapter 8

1. (a) Notice sur le *Conservatoire Impérial des Arts et Métiers, De l'enquête sur l'enseignement professionel*, 1865, *2*, 2; (b) Tresse, E.: *Revue d'histoire des sciences*, 1957, *10*, 167–174; (c) *Revue de l'enseignement technique*, 1965, 186–187.
2. Procès-Verbaux, *Conseil de Perfectionnement, Conservatoire des Arts et Métiers*, henceforth referred to as *PVCP*, 17 Jan., 1845.
3. *PVCP*, 24 Oct., 1856.
4. The 'short list' consisted of Bella, Boussingault, Dezeimeris, Lecouteux, Lefèvre, and Philiparr.
5. *PVCP*, 191–192.
6. *PVCP*, 24 Mar., 1845.
7. *C. R.*, 1845, *20*, 1371 and 1414.
8. *C. R.*, 1845, *21*, 515–518. Boussingault's modification of the safety test using ether may have been an improvement on the former method using a mixture of hydrogen and air, but in a pit explosion at Haswell, Durham, in the previous year it was claimed that the flame of the safety lamp would burn outside the lamp if there was a draught, which there certainly was in well-ventilated mines. (Berman, M.: 1976, *Social Change and Scientific Organization: The Royal Institution 1799–1844*, 174–181.)
9. *PVCP*, 4 Sept., 1845.
10. *A. S. B.* Boussingault to Dumas, 27 Dec., 1845.
11. *PVCP*, 7 Nov., 1845.

12. Boussingault, J. B.: 1858, *La fosse à fumier*, leçon professée au *Conservatoire Impérial des Arts et Métiers*.
13. Leroux, D.: 1936, 'J. B. Boussingault créateur de la chimie agricole . . .', *Bulletin de l'Association française pour l'étude du sol, 2*, 251.
14. *Revue scient.*, 1898, *10*, 70.
15. *A. S. B.*, Boussingault to Dumas, 29 Aug., 1846.
16. See Note 35 in Chap. 7.
17. Sources for the history of Liebfrauenberg are: 'Le Liebfrauenberg', *Les Vosges*, Bulletin officiel du Club Vosgien, 1957, No.4, 11–15; Pfleger, A.: 'J. B. Boussingault und Schloss Liebfrauenberg', *Elsassland*, 1920, *1*, 390–393; Chambrier, P. de: 1919, *Historique de Pechelbronn 1498–1918* (Neuchâtel), 47.
18. Wackermann, G.: 1951, *Le champ pétrolifière de Pechelbronn*, Travaux du laboratoire de Géographie de l'Université de Strasbourg, 93.
19. Chambrier, see Note 17(c), 31.
20. *PVCP*, 14 Oct., 1869.
21. *PVCP*, 19 Nov., 1850.

Chapter 9

1. *Mém. 2*, 270. Mother to Boussingault, 24 June, 1820.
2. *PVCP*, 8 Mar., 1849 – Ville was replaced by Auguste Houzeau, afterwards professor of agriculture at Rouen.
3. Roux, Eugène: 1944, *Souvenirs sur Georges Ville*, Société française d'histoire de l'agriculture, presented at a meeting on 26 June, 1944. Roux was assistant to Ville at the *Muséum d'histoire naturelle* 1894–1897. A copy of this duplicated lecture is at the Muséum.
4. *PVCP*, 11 Dec., 1851.
5. Bib. Centrale du Muséum d'Histoire Naturelle, MS 2642 letter 68. Ville to Boussingault, undated, but read at a meeting of the Conseil de Perfectionnement, 8 Mar., 1849.
6. *PVCP*, 25 Jan., 1849.
7. *A. S. B.* Boussingault to Dumas, 14 July, 1872.
8. Marx, K.: 1875 *Critique of the Gotha Programme* (Martin Lawrence, 1933), 87.
9. Archives du Bas-Rhin, Strasbourg, 2M 73. Boussingault was admitted to the Assemblée Nationale, 4 May, 1848 (*C. R. des séances de l'Assemblée Nationale*, 1848, *1*, 5).
10. Robert, Ad., Bourloton, E. and Cougny, G.: 1891, *Dictionnaire des Parlementaires français, 1*..
11. *Le Moniteur universelle*, 7 Oct., 1848.
12. *C. R. des séances de l'Assemblée nationale*, 1848, *4*, 537–564.
13. *Ibid.*, 1849, *10*, 338.
14. *Ibid.*, 564.
15. Freedemann, C. E.: 1961, *The Conseil d'Etat in Modern France* (New York), 6–7.
16. *C. R. des séances de l'Assemblée nationale*, 1849, *10*, 57.
17. Salet, G.: 'J. B. Boussingault', *Revue scient.*, 1888, *41*, 373.
18. Aucoc, L.: 1876, *Le Conseil d'Etat*, 123–131. This work has supplied most of the

information on the *Conseil d'Etat* the library of which is now accommodated in the Palais Royal.

19. (a) *PVCP*, 19 Mar., 1850; (b) *Ibid.*, 19 Nov., 1850.
20. *Ibid.*, 11 Dec., 1851.
21. *Ibid.*, 8 Jan., 1852.
22. *Annls Inst. natn. agron. Paris, 1876–1901*, 2e serie, *1*, 5.
23. Bib. du Muséum d'Histoire naturelle MS 2642, Archbishop of Paris to Ville, 1 Jan., 1852.
24. *Académie des Sciences*, papers of J. B. Dumas, carton 16.
25. Gay-Lussac and Thenard, L. J.: 1811, *Recherches physico-chimiques.*
26. *A. C.*, 1852, *35*, 5–54.
27. *A. C.*, 1880, *19*, 464–472.
28. English patent 1416 of 1880 taken out by the Brin brothers.
29. Collection, Mme F. Boussingault, Les Montils.
30. Salet, see Note 17.

Chapter 10

1. Berthollet, C. L.: *Mémoires de l'Académie Royale des Sciences*, 1785, 351–349.
2. Fourcroy, A. F.: *Annales de chimie*, 1789, *1*, 40–46.
3. Chaptal, J. A.: 1840, *Chemistry Applied to Agriculture* (New York), 58, trans. *Chimie appliquée à l'Agriculture*, 1829.
4. Liebig, J. von: 1840, *Chemistry in Its Applications to Agriculture and Physiology*, 73.
5. (a) *A. C.*, 1838, *67*, 16, in which Hermbstaedt's theory is mentioned; (b) *A. C.*, 1838, *69*, 353–367.
6. Ville, G.: *C. R.*, 1850, *31*, 578–580.
7. A. C., 1854, *41*, 5–60.
8. Ville, G.: *C. R.*, 1854, *38* 705–709; 723–727.
9. *C. R.*, 1854, *38*, 717.
10. *Ibid.*, 718.
11. 'Absorption de l'azote de l'air par les plantes', Extrait des *C. R. officials de l'Académie des Sciences*, 1855.
12. A. S. B., Boussingault to Peligot, undated but about Oct.–Nov., 1854.
13. A. S. B., Boussingault to Dumas, 23 June, 1854.
14. A. S. B., Boussingault to Peligot, 18 Oct., 1854.
15. A. S. B., Boussingault to Dumas, undated, 1855.
16. *A. C.*, 1855, *43*, 149–223, 225.
17. Dennison, J. E.: *Jl R. agric. Soc.*, 1856, *17*, 58.
18. Wells, D. A. (ed.): *Annual of Scientific Discovery for 1856* (Boston), 272.
19. Jodin, F. V.: *C. R.*, 1862, *55*, 612–615.
20. Winogradsky, S. W.: *Annales de l'Institut Pasteur*, 1890, *4*, 213.
21. Hellreigel, H. and Wilfarth, H., cited by McKee, H.S., 1962: *Nitrogen Metaboliem in Plants* (Oxford), 107.
22. Lachmann, J., cited by Russell, E. J.: 1912, *Soil Conditions and Plant Growth*, Chap.1 in all editions.

23. Ville, G.: *Recherches expérimentales sur la végétation*, 1857 ed., 55.
24. Lawes, J. B., Gilbert, J. and Pugh, E.: *Phil. Trans. R. Soc.*, 1861, *151*, 431–578.
25. *Ibid.*, 541.
26. Davy, H.: 1815, *Elements of Agricultural Chemistry*.
27. Daubeny, C. G.: *Phil. Trans. R. Soc.*, 1845, 179–253, cited by Russell, see Note 22, p. 88.
28. Lawes, J. B. and Gilbert, J. H.: 1854, *Rep. Br. Ass. Advmt Sci.* Notices and Abstracts, 70–71.
29. Berzelius, J.: 1832, *Traité de chimie* (trans. Esslinger), *6*, 570.
30. Davy, see Note 26, Chap 10, 308.
31. Mulder, G. J.: 1849, *Chemistry of Vegetable and Animal Physiology*, (Trans. Fromberg, P.), 147.
32. Sprengel, C.: 'Über pflanzenhumus, Humussäure and humussäure Salze', *Archiv für die gesammte Naturlehre*, 1826, *8*, 177, cited by Browne, C.: 1944, *Chronica Botanica* (Waltham), *8*, 177.
33. Howard, A.: 1940, *An Agricultural Testament*, 182.
34. Wiegmann, A. F. and Polstorff, L.: 1842, *Über die anorganischen Bestandtheile de Planten* (Brunswick), cited by Browne, see Note 32, 221.
35. Borchardt, A.: *C. R.*, 1843, *17*, 112–120.
36. Kuhlmann, C. F.: *C. R.*, 1843, *17*, 1118–1128.
37. Kuhlmann C. F.: *A. C.*, 1846, *18*, 145.
38. (a) *C. R.*, 1855, *41*, 845–857; (b) *A. C.*, 1856, *46*, 5–41.
39. A. S. B. Boussingault to Dumas, 8 Sept., 1855.
40. *C. R.*, 1857, *44*, 940–953.
41. Ville, G.: *C. R.*, 1857, *45*, 996–999.
42. Roux, Eugene: 1944, *Souvenirs sur Georges Ville*, see Note 3 in Chap. 9.
43. *Ibid.*
44. Bib. Centrale du Muséum d'Histoire Naturelle MS 2642. Letter 171, Ville to Larombière, 20 Sept., 1889; letter 206, Ville to Dr V. Malespire, undated.
45. *Economie rurale*, 2nd ed. 1851, *2*, 99.

Chapter 11

1. Longchamp, *A. C.*, 1826, *33*, 5–29.
2. Liebig, J. von.: *A. C.*, 1827, *35*, 329–333.
3. The acrimonious discussion between Longchamp and Gay-Lussac was continued in *A. C.*, 1827, *34*, 86–95 and 215–220.
4. *C. R.*, 1857, *44*, 108–119.
5. *C. R.*, 1859, *48*, 657–667.
6. *Ibid.*, 664.
7. Liebig, J. von.: 1840, *Chemistry in Its Applications to Agriculture and Physiology*, 5th ed. 1846, 368.
8. *Rural Economy*, 1845 (trans. G. Law), 297.
9. *C. R.*, 1859, *48*, 931–939.
10. *Ibid.*, 939.
11. *A. C.*, 1855, *43*, 191.

12. Bineau, A.: *A. C.*, 1856, *46*, 60–66.
13. *A. C.,* 1861, *61*, 363–367.
14. *Ibid.*, 363.
15. *Agronomie, Chimie Agricole et Physiologie*, 1866–1891, 8 vols, 3rd ed. *1*, 301–302.
16. Scheele, C. W.: 1785, *Mémoires de chimie* (Dijon and Paris), 51–68.
17. Pasteur, L.: (a) *C. R.*, 1857, *45*, 913–916; (b) *C. R.*, 1859, *48*, 337–338.
18. See Note 13.
19. Frazier, W. C.: 1958, *Food Microbiology* (New York), 302–303.
20. See Note 13, 365.
21. *A. C.*, 1873, *29*, 186–206.
22. Liebig, J. von: see Not 7, 1840 ed., 280.
23. Descroizilles, F. A. H.: 1795, *J. Arts Mfrs 1*, 256; but previously mentioned as Descroizelles' method by Berthollet, C. L.: 1789, *Annales de chimie 2*, 151.
24. Lenglen, M.: 1957, *Boussingault à la Société Centrale d'Agriculture* (Beauvais), 24–25.
25. See Note 21, 180.
26. *Ibid.*, 187.
27. *A. C.*, 1876, *8*, 5–24.
28. Way, J. T.: *Jl R. agric. Soc.* 1856, *17*, 125–162.
29. *Ibid.*, 147–148.
30. Muller, A., cited by McKee, H. S.: 1962, *Nitrogen Metabolism in Plants* (Oxford), 107.
31. Schloesing, T. and A. Muntz: *C. R.*, 1877, *84*, 301 and *85,* 1018.
32. Warington, R.: 1878, *J. Chem. Soc. 33*, 44.
33. Soyka, J.: *Z. Biol.*, 1878, *14*, 449: cited by McKee, see Note 30.
34. Winogradsky, S. W.: 1890, *Ann. Inst. Pasteur 4*, 213.
35. *C. R.*, 1860, *51*, 711.
36. Pasteur, L.: *C. R.*, 1862, *54*, 265–270.
37. Liebig, J. von: *Familiar Letters on Chemistry* . . . , ed. W. Gregory, 3rd ed., 1851, 210 and 237.

Chapter 12

1. *PVCP*, 28 Oct., 1853.
2. *PVCP*, 27 Nov., 1853.
3. *PVCP*, 7 Dec., 1853.
4. See Appendix 1.
5. *PVCP*, 26 Dec., 1855.
6. *PVCP*, 24 Oct., 1856.
7. *PVCP*, 15 Oct., 1859.
8. *PVCP*, 23 Jan., 1864.
9. *PVCP*, 2 Feb., 1864.
10. *Notice sur les titres et travaux de M. Eugène Tisserand*, 1883, 6.
11. Demolon, A.: 1946, *L'évolution scientifique et l'agriculture française*, 39 and 49.
12. *Annls Inst. natn. agron. Paris, 1876–1901*, 2e serie, *1*, Chap. 1. This was a special

volume describing the work of the Institut for this period, but the first chapter is devoted to the original Institut of 1850.

13. *Ibid.*, 63.
14. *Ibid.*, 12.
15. Lenglen, M.: 1937, *J. B. Boussingault à la Société Centrale d'Agriculture* (Beauvais), III.
16. *Ibid.*, 112.
17. *Ibid.*, 111.
18. *Economie rurale*, 1851, 2nd ed., *I*, viii.
19. *PVCP*, 19 Jan., 1860.
20. *PVCP*, 2 Feb., 1860.
21. Troost, L. J.: in *Funérailles de M. Boussingault*, 1887.
22. Bib. du Conseil d'Etat, No. 15429 of 25 July, 1851.
23. *Rapport Généraux des Travaux du Conseil d'Hygiène* (*Dépt. de la Seine*), 1840, 17.
24. *Ibid.*, 1843, 227.
25. Lenglen, M.: see Note 15, 115.
26. *Ibid.*, 116–117.
27. *Ibid.*, 117–120.
28. *Rapport Généraux des Travaux du Conseil d'Hygiène* (*Dépt. de la Seine*), 1870, 52.
29. *A. C.*, 1875, *4*, 285–291.
30. *A. S. B.* Boussingault to Dumas, 10 Sept., 1874.
31. Boussingault, Joseph: *A. C.*, 1866, *8*, 210–256; 1867, *11*, 434–452; 1872, *26*, 362–376.
32. *C. R.*, 1857, *44*, 116.
33. Grouvelle, P.: 1845, *Chauffage et ventilation de la Nouvelle Force à Paris.*
34. *C. R.*, 1843, *17*, 68.
35. 'Resumé du mémoire sur les lichens en réponse aux questions posées par MM. Guinon, Marnas et Bonnet', in *Opinion des chimistes*, 1861; 'Observations sur l'avis donné par M. Salvetat en réponse aux questions posées par la Compagnie Parisienne', in *Opinion des chimistes*, 1865. These pamphlets are in the Bibliothèque Nationale.

Chapter 13

1. Malpighi, M.: 1687, *Anatome plantarum 1*, 37.
2. Mariotte, E.: 1717, *Oeuvres de M. Mariotte* (Leiden), 135.
3. Hales, S.: 1727, *Vegetable Staticks*, Scientific Book Guild edition 1961, 185–186.
4. Priestley, J.: 1775, *Experiments and Observations on Different Kinds of Air 1.*
5. Berthollet, C. L.: 1785, *Observations sur la physique, sur l'histoire naturelle et sur les arts*, 26.
6. Ingenhousz, J.: 1779, *Experiments upon Vegetables* . . . , 38–44.
7. Senebier, J.: 1782, *Mémoires physico-chimiques* (Geneva).
8. De Saussure, N. T.: 1804, *Recherches chimiques sur la végétation* 49–55; 217–239.
9. Pelletier, F. J. and Caventou, J. B.: *A. C.*, 1818, *9*, 195.

10. Dutrochet, H.: 1837, *Mémoires pour servir à l'histoire anatomique et physiologique des végétaux et des animaux 1*, 365 and 431.
11. Mohl, H. V., cited by Green, J. R.: 1909, *A History of Botany 1860–1900*, 260 and 279.
12. Meyer, J. R.: 1845, *Die Organische Bewegung in ihren Zusammenhang mit dem Stoffwechsel* (Heidelberg), cited by Palladin, V. I.: 1923, *Plant Physiology* (Philadelphia), 32.
13. *A. S. B.* Boussingault to Dumas, 23 June, 1854.
14. de Saussure, see Note 8, 57.
15. Lacroix, A.: 1932, *Figures de Savants 2*, Plate 7, part of a letter from Boussingault to Peligot, 20 Aug., 1856.
16. *A. C.*, 1862, *66*, 298–429.
17. Green, J. R., see Note 11, 281–282.
18. *A. C.*, 1868, *13*, 219–240.
19. *Ibid.*, 237.
20. *A. C.*, 1868, *13*, 282–416.
21. Duchartre, P.: *C. R.*, 1866, *63*, 854–856.
22. Barthélemy, A.: *C. R.*, 1877, *84*, 663–666.
23. Sachs, J. von: 1865, *Handbuch der Experimental-Physiologie des Pflanzen* (Leipzig); *Lehrbuch der Botanik* (Leipzig) (trans. Bennet, A. and Dyer, N. T.: 1975 as *Textbook of Botany morphological and physiological* (Oxford), 75, 86–87, 603).
24. Barthélemy, see Note 22, 663.
25. Merget, E.: *C. R.*, 1878, *86*, 1492–1495.
26. Blackman, F. F.: *Phil. Trans. R. Soc.*, 1895, *B186*, 485–562.
27. *A. C.*, 1869, *18*, 321–352.
28. De Saussure, see Note 8, 54.
29. van Thiegham, *C. R.*, 1867, *65*, 867–871.
30. Brown, N. T. and Esconbe, F.: 1905, *Proc. Roy. Soc. Lond., B76*, 29.
31. Boussingault, see Note 27, 336–338.
32. *Ibid.*
33. Chevreul, M. E.: *Mémoires de l'Académie des Sciences, Institut de France*, 1861, *33*. This mémoire occupied about a thousand pages.
34. *A. C.*, 1876, *8*, 433–443.
35. *Ibid.*, 439.
36. Pasteur, L.: *C. R.*, 1876, *82*, 792–793. This was Pasteur's reply to Boussingault's mémoire (Note 34).
37. Boussingault, see Note 34, 441.
38. This analogy is taken from Rabinowitch, R. I. and Govindjie, 1969, *Photosynthesis* (New York), 58

Chapter 14

1. *Report of the Commission for the Exhibition of 1862* (HMSO), 1863, 154 and 157.
2. *Rapport de l'administration de la Commission Impériale sur la section française de l'Exposition Universelle de 1862*, 1864, 48–49.

3. Boussingault, Edmund: *Document sur l'expédition du Mexique. Conversation de J. B. Boussingault avec l'Empereur Napoleon III, 6 juin 1862* (Antibes 1927). Edmund, a grandson of Boussingault, was a son of Joseph Boussingault.
4 *Ibid.*, 13.
5. *Les établissements Jacob Holtzer*, publisher Devambes, no place of publication stated, date probably 1929, p. 28; a copy at the Château Holtzer, Unieux.
6. *Archives de la Commission Scientifique du Mexique*, 1865, *I*, 8.
7. *Ibid.*, 209–224.
8. *Ibid.*, 1867, *3*, 10, 11.
9. *A. S. B.* Boussingault to Dumas, 10 Jan., 1866.
10. Boussingault, Edmund, Note 3, 14.
11. *PVCP*, 16 Oct., 1860.
12. *PVCP*, 14 Oct., 1869.
13. *Etat nominatif des habitants – commune de Lampertsloch et commune de Goersdorf*, 1836 and 1866. Archives de Bas-Rhin, Strasbourg, MS 7M 489, MS 7M 389.
14. *Les établissements Jacob Holtzer*, Note 5, 29.
15. *C. R.*, 1861, *52*, 1008–1011.
16. Bouis, J.: *C. R.*, 1861, *52*, 1195.
17. *C. R.*, 1861, *52*, 1249–1252.
18. Lavoisier, A. L.: *Observations Physiques*, 1784, *24*, 368–380; cited by Partington, J. R.: 1964, *A History of Chemistry*, *3*, 446.
19. Boussingault, Note 15, 1010.
20. Boussingault, Note 17, 1250.
21. *C. R.*, 1861, *53*, 5–10. In this paper Boussingault mentions the assistance of Caron and Bouis.
22. *C. R.*, 1868, *66*, 873–877.
23. (a) Hadfield, R. A.: 1925, *Metallurgy and Its Influence on Modern Progress*, 72; (b) Austen-Roberts, W. C.: 1890, *An Introduction to the Study of Metallurgy*, 4th ed., 1898, 37.
24. (a) Singer, C., Holmyard, E. J., Hall, A. R. and Williams, T. I.: 1957, *A History of Technology* (Oxford), *3*, 32–34; (b) Mellor, J. W.: *A Comprehensive Treatise on Inorganic and Theoretical Chemistry*, 1957 imp. *12*, 737.
25. Partington, J. R.: 1964, *A History of Chemistry*, 3, 510, citing *Obs. Phys.*, 1786, *29*, 210, 281.
26. Cementation theories are discussed in Mellor (see Note 24b), *12*, 736–744.
27. *A. C.*, 1875, *5*, 145–265. This paper on the transformation of iron into steel by cementation was published in book form, also in 1875.
28. *Ibid.*, 145–147.
29. *Ibid.*, 218–247.
30. Partington, see Note 25, *3*, 510.
31. *Ibid.*, 530, citing article 'Acier' in *Encyclopédie Méthodique, 1786*.
32. Boussingault, see Note 27, 225.
33. More detailed information on steel formation is to be found in Alexander, W. and Street, A.: 1944, *Metals in the Service of Man*, Chap. 11, 4th ed., 1962.
34. Boussingault, see Note 27. 148–167.
35. Margueritte, F.: *A. C.*, 1846, *18*, 244.

36. Mohr, C. F.: 1870, *Lehrbuch der chemisch-analytischen Titrimethode* (Braunschweig), 169, cited by Mellor, Note 24b, *14*, 288.
37. *A. C.*, 1876, *8*, 145–175.
38. *A. C.*, 1878, *15*, 99–100.
39. Singer, C. et al., see Note 24a, *5*, 64.
40. Stodart, J. and Faraday, M.: *Phil. Trans. R. Soc.* 1822, *112*, 253.
41. Stodart and Faraday, *Quarterly Journal of Science, 9*, 319.
42. (a) Hadfield, see Note 23a, 71; (b) Mellor, see Note 24b, *13*, 586, who cites Berthier, P. 'Sur les alliages du chrome avec le fer et avec l'acier', *Traité des essais par la voie sèche*, 1834, *2*, 475; (c) Partington, see Note 25, *4*, 98, cites Berthier's original paper as *A. C.*, 1821, *17*, 35.
43. Carr, J. C. and Taplin, W.: 1962, *History of the British Steel Industry* (Oxford), 132.
44. U.S. Patent 49495 of 1865, cited by Mellor, see Note 42b.
45. Boussingault, see Note 38, 93.
46. Hadfield, see Note 23a; Austen-Roberts, see Note 23b.
47. (a) *A. S. B.*, Boussingault to Dumas, 14 July, 1872; (b) *A. C.*, 1872, *27*, 477–503; *C. R.*, 1872, *75*, 229–232.
48. *PVCP*, 18 Dec. 1866.
49. *PVCP*, 12 July, 1868. The excessively long interval between this meeting and that of 18 Dec., 1866 is not explained.
50. *PVCP*, 13 Apr., 1869.
51. *PVCP*, 14 Oct., 1869.

Chapter 15

1. Fohlen, C.: 1964–65, *Histoire de Besançon, 1*, 252.
2. *A. S. B.* Boussingault to Dumas, 26 Feb., 1870.
3. *A. S. B.* Boussingault to Dumas, 11 Mar., 1870.
4. Académie des Sciences, J. B. Dumas dossier, 'Commission des Monnaies 1867–70'.
5. *A. S. B.*, Boussingault to Dumas, 15 Sept., 1875.
6. Pasteur, L.: *Bulletin des Séances de la Société nationale d'Agriculture Français*, 1880, *40*, 815–819.
7. Obituary notice in the *Journal of the Chemical Society*, 1888, *53*, 512.
8. Partington, J. R.: 1962, *A History of Chemistry, 4*, 396.
9. Collection Mme F. Boussingault, Les Montils, Boussingault to Joseph B., 23 Aug., 1870.
10. *PVCP*, 16 Oct., 1871.
11. Howard, M.: 1961, *The Frnaco-Prussian War*, 13–14.
12. Note 9, Boussingault to Joseph B., 12 Aug., 1870.
13. *Ibid.*, Boussingault to Joseph B., 23 Aug., 1870.
14. *Ibid.*, Boussingault to Joseph B., 19 Aug., 1870.
15. *Ibid.*, see Note 12.
16. Crosland, M.: 1976, 'Science and the Franco-Prussian War', *Social Studies of Science 6*, 195.

17. *C. R.*, 1870, *71*, 728, cited Crosland, *ibid.*, 199.
18. *A. C.*, 1839, *71*, 65–79; *C. R.*, 1838, 7, 1019–1020.
19. *PVCP*, 7 Feb., 1872.
20. *Bull. Seanc. Soc. nat. Agric. Fr.*, 1871–72, 7, 300.
21. *Inauguration du monument de Boussingault au Conservatoire des Arts et Métiers à Paris le Dimanche 7 juillet 1895*, 27.
22. *A. S. B.* Boussingault to Dumas, 14 July, 1872.
23. *Bull. Soc. Encour. Ind. natn. Paris*, 1872, *19*, 219–234.
24. Note 9, Adèle Boussingault to Benoît Fourneyron, 6 June, 1858.
25. *A. S. B.* Boussingault to Dumas, 15 Sept., 1875.
26. Note 9, Adèle Boussingault to Joseph B., 24 Apr., 1877, possibly her last letter as it has a note attached to it in Joseph's handwriting saying that his mother fell ill of '*mal de forêt*' and that she had diabetes.
27. *A. S. B.* Boussingault to Dumas, 10 Mar., 1874.
28. *A. S. B.* Boussingault to Dumas, 8 May, 1875 and 15 Sept., 1878.
29. *A. S. B.* Boussingault to Dumas, 8 May, 1875.
30. Dumas, see Note 4.
31. *A. S. B.* Boussingault to Dumas, 26 Jan., 1881.
32. See Note 9, Adèle Boussingault to Joseph B., undated but possibly 1872, from Liebfrauenberg.
33. Bib. Chateau Holtzer, Unieux, correspondence.
34. *Ibid.*, Brustlein to Berthe Holtzer, 17 Nov., no year stated but probably 1880.
35. Bib. Chateau Holtzer, Unieux, Paul Duthu (technical director Ria steelworks, Pyrénées Orientales) to Berthe Holtzer, 27 Aug., 1884.
36. See Note 9. Newspaper cuttings of Boussingault's funeral of which the best is from *La Justice* of 14 May, 1887.
37. Chambrier, P. de: 1920, *Les Mines de Pétrole de Pechelbronn* (Strasbourg), 9.
38. See Note 9, Tribunal arbitral mixte Franco-Allemand, Affaire Boussingault c/Etat Allemand s. 16 – 1088, requested 25 Feb., 1924.
39. Association des amis de la maison de l'église 'la Liebfrauenberg', Strasbourg, *Liebfrauenberg* and *Liebfrauenberg la maison de l'église*.

Chapter 16

1. Lundgren, P.: 1980, 'The organization of science and technology in France: a German perspective', Fox, R. and G. Weisz, *The organization of science and technology in France 1800–1914*, 321–323.
2. Zwerling, C.: 1980, 'The emergence of the Ecole Nationale Supérieure', Fox, R. and G. Weisz, *Ibid.*, 44–49.
3. Ville, G.: 1869, *Les engrais chimiques*, trans. W. Crookes, 1879, *On Artificial Manures*, 2nd ed., *Artificial Manures* (1909).
4. Berman, M.: 1978, *Social change and scientific organization: The Royal Institution 1799–1844*, 49.
5. Berman, *Ibid.*, 54.
6. *Inauguration du monument de Boussingault au Conservatoire des Arts et Métiers à Paris le Dimanche 7 juillet 1895*, discours de M. Ach. Müntz.

7. Prjanischnikow, D. N.: 1936, *Bulletin de l'Association français pour l'étude du sol, 2*, 167–173, trans. from *Izvestia* of 30 May, 1936.
8. Ostwald, W.: 1909, *Grosse Männer* (Leipzig), trans. M. Dufour, 1918, *Les Grands Hommes*.
9. Kisker, G. N.: 1964, *The Disorganized Personality*, 223 (New York), has proved useful in making a general assessment of Boussingault's personality although it was far from being 'disorganised'.
10. Morrell, J. B.: 1972, 'The Chemist Breeders: The Research Schools of Liebig and Thomas Thomson', *Ambix 19*, No. 1, 1–7.
11. Collection Mme F. Boussingault, Les Montils, J. H. Gilbert to Boussingault, 27th April, 1876.
12. Ville, *op. cit.*, Note 3.
13. Deherein, P. P.: 1887, 'L'oeuvre agricole de M. Boussingault', *Revue scient. 2*, 37.
14. *C. R.*, 1872, *74*, 87–90.
15. Beattie, J.: 1964, *Other Cultures*, 42.
16. The concept of a 'paradigm' in the philosophy of science was introduced by Thomas Kuhn in his *Structure of Scientific Revolutions* (Chicago, 1970). Various interpretations of the term have been claimed by his critics but an elementary example will will suffice for our purpose. According to Kuhn, competing theories are temporarily resolved in a paradigm which includes those which are acceptable by the scientific community, and excluding those which are not acceptable. Facts and conclusions may later arise which challenge the validity of the paradigm requiring some adjustment to the paradigm; or a crisis point is reached when the existing paradigm must be replaced by another.
17. Anschutz, R.: 1929, *August Kekulé*, 1, 672 (Berlin). This volume includes Appendix 9, the minutes of the Karlsruhe Congree, published in French by Charles Wurtz, 671–688.
18. *Ibid.*, 677.
19. *Ibid.*, 688.
20. Cannizzaro, S.: 1858, *Il Nuova Cimenta 7*, 321–366 (Genoa), trans. as *Sketch of a Course of Chemical Philosophy*, Alembic Club reprint No. 18 (Edinburgh, 1910).

BOUSSINGAULT'S SCIENTIFIC PAPERS

(a) 1821–1835

South America

Sur le chica que les Indiens emploient pour se peindre le corps en rouge.	*A. C.*, 1824, *27*, 315–323.
Sur les propriétés chimiques du rocou.	*A. C.*, 1825, *28*, 440–444.
Observations sur quelques carbonates.	*A. C.*, 1825, *29*, 283–288.
Note sur la cera de palma que l'on recueille dans les Andes de Quindiu.	*A. C.*, 1825, *29*, 330–333.
Sur l'existence de l'iode dans l'eau d'une saline de la province d'Antioquia.	*A. C.*, 1825, *30*, 91–96.
Analyse de l'alumine sulfatée native du Rio-Saldana.	*A. C.*, 1825, *30*, 109–111.
Analyse d'un calcul urinaire de nature ferrugineuse.	*J. P.*, 1825, *11*, 153–156.
Analyse d'une nouvelle substance minérale (gay-lussite).	*A. C.*, 1826, *31*, 270–276.
Sur le gisement du platine.	*A. C.*, 1826, *32*, 204–212. *A. M.*, 1827, *1*, 175–178.
Mémoire sur une nouvelle méthode pour l'essai et le traitement de la pyrite aurifère.	*A. C.*, 1827, *34*, 253–268. *A. M.*, 1827, *1*, 319–333.
Sur la composition de l'or natif argentifère.	*A. C.*, 1827, *34*, 408–419.
Analyse de l'arsèniate de fer de Loaysa, près Marmato, province de Popayan.	*A. C.*, 1829, *41*, 75–78.
Analyse de la blende noire de Marmato, province de Popayan.	*A. C.*, 1830, *43*, 312–316.
Note sur la présence de l'ammoniaque dans l'oxide de fer naturel.	*A. C.*, 1830, *43*, 334–335.
Analyse d'un nouveau minéral trouvé dans le Paramo-Ricco, près Pamplona.	*A. C.*, 1830, *45*, 325–329.
Analyse de l'eau minérale de Paipa, près Tunja.	*A. C.*, 1830, *45*, 329–324.
Analyse de différentes variétés d'or natif.	*A. C.*, 1830, *45*, 440–443.
Recherches sur la cause qui produit le goître dans les Cordilières de la Nouvelle-Grenade.	*A. C.*, 1831; *48*, 41–69.
Analyse de l'eau du Rio-Vinagre.	*A. C.*, 1832, *51*, 107–110.

Recherches sur les phénomènes chimiques qui se passent dans l'amalgamation américaine.	*A. C.*, 1832, *51*, 337–356.
Recherches chimiques sur la nature des fluides élastiques qui se dégorgent des volcans de l'équateur.	*A. C.*, 1833, *52*, 5–23.
Considérations sur les eaux thermales des Cordilières.	*A. C.*, 1833, *52*, 181–190.
Observations sur la rayonnement nocturne, faites dans les Cordilières de la Nouvelle-Grenade.	*A. C.*, 1833, *52*, 260–266.
Analyse de l'alumine sulfatée du volcan de Pasto.	*A. C.*, 1833, *52*, 348–351.
Examen chimique d'une substance minérale deposée par l'eau chaude de Coconuco, près Popayan.	*A. C.*, 1835, *52*, 396–399.
Mémoire sur la profondeur à laquelle se trouve la couche de température invariable entre les tropiques. Détermination de la température moyenne de la zone torride au niveau de la mer. Observations de la décroissement de la chaleur dans les Cordilières.	*A. C.*, 1833, *53*, 225–247. *C. R.*, 1835, *1*, 385–387.
Analyse de l'halloysite de Guatéqué dans la Nouvelle-Grenade.	*A. C.*, 1833, *53*, 439–441.
Examen d'une substance considerée comme un composé d'hydrogène et de platine.	*A. C.*, 1833, *53*, 441–444.
Mémoire sur les salines iodifères des Andes.	*A. C.*, 1833, *54*, 163–178.
Mémoire sur l'action du gaz acide hydrochlorique sur l'argent à une haute température; observation sur le départ sec.	*A. C.*, 1833, *54*, 253–263.
Sur le sous-oxide de plomb.	*A. C.*, 1833, *54*, 264–266.
Recherches sur la nature de l'acide phosphorique qui constitue les phosphates naturels.	*A. C.*, 1833, *55*, 185–192.
Sur la composition du vernis des Indiens de Pasto.	*A. C.*, 1834, *56*, 216–221.
Recherches sur la composition de l'atmosphère; sur la possibilité de constater l'existence des miasmes; sur la présence d'un principe hydrogéné dans l'air.	*A. C.*, 1834, *57*, 148–182.
Sur les tremblements de terre des Andes.	*A. C.*, 1835, *58*, 81–88. *B. G. F.*, 1834–35, *6*, 52–57.
Ascension au Chimborazo, exécutée le 16 décembre 1831.	*A. C.*, 1835, *58*, 150–180.
Sur la composition de la cire de palmier.	*A. C.*, 1835, *59*, 19–24.

Boussingault and Rivero:

Mémoire sur le lait de l'arbre de la vache.	*A. C.*, 1823, *23*, 219–224.
Sur les eaux chaudes de la Cordilière de Venezuela.	*A. C.*, 1823, *23*, 272–276.
Résultats des observations barométriques faites à la Guayra.	*A. C.*, 1824, *25*, 427–429.
Mémoire sur différentes masses de fer qui ont été trouvées sur la Cordilière des Andes.	*A. C.*, 1824, *25*, 438–443.
Mémoire sur le lait vénéneux de l'ura crépitans.	*A. C.*, 1825, *28*, 430–435.
Mémoire sur l'urao (carbonate de soude).	*A. C.*, 1825, *29*, 110–111.
Observations du baromètre faites à Bogotá.	*A. C.*, 1823, *23*, 427.
Observations barométriques, faites à Santa-Fé-de-Bogotá.	*A. C.*, 1827, *34*, 203–214.

Boussingault and Roulin:

Examen chimique du curare, poison des Indiens de l'Orénoque.	*A. C.*, 1828, *39*, 24–37.

Metallurgy

Note sur la combinaison du silicium avec le platine, et sur sa présence dans l'acier	*A. C.*, 1821, *16*, 5–15.

Chemistry

Note sur le sel ammoniaque que produit une mine de houille incendiée, par les Elèves mineurs de Saint-Etienne.	*A. C.*, 1822, *21*, 158–159. *A. M.*, 1823, *8*, 304–305.
Sur la composition de l'acide pyro-mucique.	*A. C.*, 1835, *58*, 106–109.

(b) 1836–1848

South America

Examen comparatif des circonstances météorologiques sous lesquelles végètent certaines plantes alimentaires à l'équateur et sous la zone temperée.	*A. C.*, 1836, *63*, 337–358. *C. R.*, 1837, *4*, 178–179.

Observations sur la pluie des Tropiques.	*A. C.*, 1836, *61*, 167–171. *C. R.*, 1836, *2*, 109–111.
Examen chimique de la banane et de la séve du bananier; suivi de considérations de sa culture et ses usages.	*C. R.*, 1836, *2*, 440–444.
Deuxieme Mémoire sur l'agriculture des tropiques; sur le cacao.	*C. R.*, 1836, *3*, 510–513.
Analyse de plusieurs échantillons d'or natif de la Nouvelle Grenade.	*C. R.*, 1837, *4*, 950–956.
Mémoire sur l'influence des défrichements dans la diminution des cours d'eau.	*A. C.*, 1837, *64*, 113–141. *C. R.*, 1837, *4*, 584–585.
M. Boussingault transmet à M. Arago des observations d'intensité magnetique, faites à Santa-Fé-de-Bogotá et sur le sommet d'une montagne voisine (Guadalupe) observations qui n'accusent aucune diminition dans l'intensité avec la hauteur – Observations diurnes de la déclinaison faites à Marmato par 5° 29′ de latitude nord.	*C. R.*, 1837, *4*, 93.
Remarques à l'occasion d'une lettre de M. de Paravey, sur l'existence de l'iode dans une source salée de l'Amérique du Sud.	*C. R.*, 1840, *10*, 79.
Sur la composition de l'air qui se trouvent dans les pores de la neige.	*A. C.*, 1841, *1*, 354–360. *C. R.*, 1841, *12*, 317–322.
Analyse d'une résine fossile des environs de Bucaramanga.	*A. C.*, 1842, *6*, 507.
Sur la rayonnement de la neige.	*C. R.*, 1842, *14*, 405–406.
Observations de quelques effets de la foudre.	*C. R.*, 1842, *14*, 835–837.
Effets produits par un coup de foudre. L'odeur qu'exhalent souvent les corps foudroyés recemment, est-elle bien désignée par le nom d'odeur sulfureuse?	*C. R.*, 1846, *22*, 919–920.
Nouvelle analyse de la gay-lussite.	*A. C.*, 1843, *7*, 488–489.
Observations sur une eau minérale acide du Paramo de Ruiz, dans la Nueva-Granada.	*A. C.*, 1847, *20*, 109–113. *C. R.*, 1847, *24*, 397–400.
M. Boussingault en presentant un Mémoire de M. J. Goudot, sur la culture de l'arracacha dans la Nouvelle-Grenade et la possibilité de l'introduire en Europe, fait remarquer que l'affection qui a frappé cette année les pommes de terre doit éveiller l'attention sur les tentatives qui auraient pour but, comme celle que provoque ce Mémoire, l'introduction d'un végétal fournissant une nourriture abondante, salubre et agréable au goût.	*C. R.*, 1845, *21*, 1002; 1149–1153. *B.S.A.*, 1845–46, 383.

Animal Husbandry

Recherches sur la quantité d'azote contenue dans les fourrages et sur leurs equivalents.	*A. C.*, 1836, *63*, 225–244. *A. C.*, 1838, *67*, 108–121. *C. R.*, 1836, *3*, 726–731. *C. R.*, 1838, *6*, 383–384.
Mémoire sur la quantité de gluten contenue dans les farines de plusieurs espèces de froment cultivées dans le même sol.	*A. C.*, 1837, *65*, 301–320. *C. R.*, 1837, *5*, 340.
Recherches sur l'influence de la nourriture des vaches, sur la quantité et la constitution chimique du lait (avec LeBel).	*A. C.*, 1839, *71*, 65–79. *C. R.*, 1838, *7*, 1019–1020.
Analyses comparées des aliments consommés et des produits rendus par une vache laitière. Recherches entreprises dans le but d'examiner si les herbivores empruntent de l'azote à l'atmosphère.	*A. C.*, 1839, *71*, 113–127. *C. R.*, 1838, *7*, 245.
Analyses comparées des aliments consommés et des produits rendus par un cheval soumis à la ration d'entretien; suite des recherches entreprisès dans le but d'examiner si les herbivores prélèvent de l'azote à l'atmosphère.	*A. C.*, 1839, *71*, 128–136. *C. R.*, 1838, *7*, 1157–1160.
Recherches sur l'engraissement des bestiaux et la formation du lait (avec Dumas et Payen).	*A. C.*, 1843, *8*, 63–114. *C. R.*, 1843, *16*, 174, 345–362.
Remarques à l'occasion d'une note de M. Liebig sur la formation de la graisse chez les animaux.	*C. R.*, 1843, *16*, 668–675.
Analyses comparées de l'aliment consommé et des excrements rendus par une tourterelle, entreprises pour rechercher s'il y a exhalation d'azote pendant la respiration des granivores.	*A. C.*, 1844, *11*, 433–456. *C. R.*, 1844, *19*, 73. *A. S. N.*, 1844, *2* (Zool.), 211–221. *B. S. A.*, 1845–46, *I*, 192–194.
Considerations sur l'alimentation des animaux.	*A. S. N.*, 1844, *2* (Zool.), 229–244.
M. Boussingault presente quelques observations relatives à la formation de la graisse chez les chevaux.	*C. R.*, 1843, *16*, 556–561.
Recherches sur la formation de la graisse chez les animaux.	*C. R.*, 1845, *20*, 1726–1728.
Recherches expérimentales sur le développement de la graisse pendant l'alimentation des animaux.	*A. C.*, 1845, *14*, 419–482.
Formation de la graisse chez les animaux.	*B. S. A.*, 1845–46, *1*, 192–194.

Recherches sur la constitution de l'urine des animaux herbivores.	*A. C.*, 1845, *15*, 97–114.
Analyse chimique des urines des animaux herbivores.	*C. R.*, 1845, *21*, 4.
Composition de l'urine des herbivores.	*J. P.*, 1846, *9*, 129–130.
Recherches sur le développement de la substance minérale dans le système osseux du porc.	*A. C.*, 1846, *16*, 486–493. *C. R.*, 1846, *22*, 356–358.
Recherches expérimentales sur la faculté nutritive des fourrages avant et après le fanage.	*A. C.*, 1846, *17*, 291–296. *C. R.*, 1846, *22*, 690.
Expériences statiques sur la digestion.	*A. C.*, 1846, *18*, 444–478. *R. C.*, 1846, *23*, 569–594.
Relation d'une expérience entreprise pour déterminer l'influence que le sel, ajouté à la ration, excerce sur le développement du betail.	*C. R.*, 1846, *23*, 949–953. *C. R.*, 1847, *24*, 636–638. *C. R.*, 1847, *25*, 729–733. *A. C.*, 1847, *19*, 117–125. *A. C.*, 1847, *20*, 113–116. *A. C.*, 1848, *22*, 116–122.
De l'emploi des fourrages trempés dans l'alimentation du betail.	*A. C.*, 1847, *19*, 249–251.
Observations sur l'influence que le sel ajouté à la ration des vaches peut exercer sur la production du lait.	*A. C.*, 1848, *22*, 503–505.
Recherches sur l'influence que certains principles alimentaires peuvent exercer sur la proportion de matières grasses contenue dans le sang.	*A. C.*, 1848, *24*, 460–463.

Crop Husbandry

Des influences météorologiques sur la culture de la vigne.	*A. C.*, 1837, *64*, 174–185. *C. R.*, 1837, *4*, 371–378.
Recherches chimiques sur la végétation, entreprises dans le but d'examiner si les plantes prennent de l'azote à l'atmosphère.	*A. C.*, 1838, *67*, 5–54. *C. R.*, 1838, *6*, 102–112, 129.
Recherches chimiques etc., Deuxième mémoire.	*A. C.*, 1838, *69*, 353–367. *C. R.*, 1838, *7*, 889–891. *A. S. N.*, 1838, *10* (Bot.), 257–268.
Recherches chimiques sur la végétation (troisième mémoire). Discussion de la valeur relative des assolements par l'analyse élémentaire.	*C. R.*, 1838, *7*, 1149–1155. *C. R.*, 1839, *8*, 54. *C. R.*, 1839, *9*, 233. *A. S. N.*, 1839, *11* (Bot.), 31–38.

	A. C., 1841, *1*, 208–246. *Mém. Acad. Sci.*, 1842, *18*, 345–384.
Mémoire sur les résidus des récoltes.	*A. C.*, 1841, *2*, 308–321. *C. R.*, 1840, *10*, 918–921.
Mémoire sur les engrais et leurs valeurs comparées (et Payen).	*A. C.*, 1841, *3*, 65–108. *C. R.*, 1841, *13*, 323–332.
Mémoire sur les engrais . . . Deuxième mémoire.	*A. C.*, 1842, *6*, 449–465. *C. R.*, 1842, *15*, 657–669.
Recherches concernant l'influence que le plus ou moins de pluie de l'année exerce sur les résultats de la culture du Madia sativa.	*C. R.*, 1842, *14*, 349–361.
Remarques à l'occasion d'un passage du Mémoire de M. Kuhlmann relatif à la fertilisation des terres par les sels ammoniacaux, par les nitrates et par d'autres composés azotes.	*C. R.*, 1843, *17*, 1153–1155.
Remarques sur l'analyse du guano de MM. Girardin et Bidard (et Payen).	*A. C.*, 1844, *10*, 237–238.
Note sur la respiration des plantes.	*C. R.*, 1844, *19*, 870; 945–948.
Expériences sur l'emploi du phosphate ammoniacomagnésien comme engrais.	*C. R.*, 1845, *21*, 722–724.
M. Boussingault présente au nom de l'auteur, M. Caillet, un tableau des observations météorologiques faites à l'Institution royale agronomique de Grignon, et, à cette occasion fait remarquer quelles ont été les différences dans les conditions météorologiques des deux années 1844 et 1845, l'une favorable, et l'autre contraire à la végétation des pommes de terre.	*C. R.*, 1846, *22*, 251.
Recherches sur le développement successif de la matière végétale dans la culture du froment.	*A. C.*, 1846, *17*, 162–171. *C. R.*, 1846, *22*, 617–618. *A. S. N.*, 1846, *6* (Bot.), 5–14.
Sur la préparation du phosphate ammoniacomagnésien au moyen de l'urine.	*A. C.*, 1847, *20*, 117–118.

Chemistry

Recherches sur la composition de l'atmosphère.	*C. R.*, 1835, *1*, 36; 87–89.
Mémoire sur l'acide subérique.	*C. R.*, 1836, *2*, 77–80.
Mémoire sur la composition des bitumes.	*C. R.*, 1836, *3*, 375–378. *A. C.*, 1837, *64*, 141–151.

Sur la composition du sucre de gélatine et de l'acide nitrosaccharique de Braconnot.	*C. R.*, 1838, *7*, 493–495. *C. R.*, 1840, *11*, 917–920. *A. C.*, 1841, *1*, 257–270.
Analyse de quelques substances bitumineuses.	*A. C.*, 1840, *73*, 442–445.
Recherches sur la véritable constiution de l'air atmosphérique (et Dumas).	*A. C.*, 1841, *3*, 257–305. *C. R.*, 1841, *12*, 1005–1025.
Recherches sur la quantité d'acide carbonique contenu dans l'air de la ville de Paris.	*C. R.*, 1841, *13*, 366. *A. C.*, 1844, *10*, 456–469.
Analyse comparative de l'air à Paris, à Berne et sur la Faulhorn.	*C. R.*, 1841, *13*, 634–636.
Observations simultanées faites à Paris et à Andilly, près Montmorency, pour rechercher la proportion d'acide carbonique contenue dans l'air atmosphèrique (et Lewy).	*A. C.*, 1844, *10*, 470–474. *C. R.*, 1844, *18*, 473–476.
Expériences constatant l'efficacité des lampes de Davy dans des mélanges d'air et de vapeurs inflammables émanant des liquides très-volatils. Emploi de la lumière produite par la pile dans les lieux où l'atmosphère est sujette à devenir détonante.	*C. R.*, 1845, *21*, 515–518. *C. R.*, 1846, *22*, 225.
Rapport sur des échantillons d'eau salée et de bitume envoyés de la Chine par M. Bertrand.	*C. R.*, 1846, *22*, 667–669.

(c) 1849–1851

South America

Rapport sur les observations météorologiques faites à l'Antisana par M. Carlos Aguirre.	*C. R.*, 1851, *32*, 741–747.

Animal Husbandry

Recherches sur la quantité d'ammoniaque contenue dans l'urine.	*A. C.*, 1850, *29*, 472–500. *J. P.*, 1850, *18*, 263–267.

Crop Husbandry

Sur la quantité de potasse enlevée au sol par la culture de la vigne.	*A. C.*, 1850, *30*, 369–371.

Chemistry

Observations sur la congélation du vin et des mélanges d'eau et d'alcool.	*A. C.*, 1849, *25*, 363–365.
Sur l'extraction du gaz oxygène de l'air atmosphérique.	*C. R.*, 1851, *32*, 261–267; 821–823.
Recherches sur le moyen d'extraire, pour les besoins des arts, le gaz oxygène contenu dans l'air atmosphérique	*A. C.*, 1852, *35*, 5–54.
Expériences ayant pour but de déterminer la cause de la transformation du pain tendre en pain raissis.	*A. C.*, 1852, *36*, 490–494. *C. R.*, 1852, *35*, 588–591.

(d) 1852–1859

South America

Sur un gisement de platine signalé dans un filon de la province d'Antioquia (Nouvelle-Grenade).	*C. R.*, 1856, *42*, 917–922.
Observations sur la hauteur du mercure dans le baromètre au niveau de la mer dans la proximité de l'equateur.	*C. R.*, 1857, *44*, 1036.
M. Boussingault présente des flèches empoisonnées par le curare et quelques observations sur la composition de ce poison.	*C. R.*, 1858, *47*, 851.
De l'emploi de la fumée pour preserver les vignes de la gelée.	*A. C.*, 1858, *52*, 485–495.
Fragments d'un Mémoire sur les gisements de guano dans les ilots et sur les côtes de l'océan Pacifique.	*C. R.*, 1860, *51*, 844–855.
Sur le gisement du nitrate de soude du Pérou	*A. C. P.*, 1861, *2*, 372–377.

Crop Husbandry

Mémoire sur la composition de l'air contenu dans la terre végétale (et Lewy).	*C. R.*, 1852, *35*, 764–775. *A. S. N.*, 1853, *19* (Bot.), 5–16. *A. C.*, 1853, *37*, 5–50.
Mémoire sur le dosage de l'ammoniaque contenue dans les eaux.	*A. C.*, 1853, *39*, 257–291. *C. R.*, 1853, *36*, 814–821. *J. P.*, 1854, *25*, 122–131.

Sur la quantité d'ammoniaque contenue dans l'eau de pluie recueillie loin de villes.	*C. R.*, 1853, *37*, 207–208; 798–806. *B. S. A.*, 1853–54, *9*, 176–177.
Mémoire sur la quantité d'ammoniaque contenue dans la pluie, la rosée, le brouillard recueillis loin des villes.	*A. C.*, 1854, *40*, 129–155. *C. R.*, 1854, *38*, 249–251. *B. S. A.*, 1853–54, *9*, 183–185.
Recherches sur la végétation, entreprises dans le but d'examiner si les plantes fixent dans leur organisme l'azote qui est à l'état gazeux dans l'atmosphère.	*A. C.*, 1854, *41*, 5–60. *C. R.*, 1854, *38*, 580–586; 607. *A. S. N.*, 1854, *1* (Bot.), 241–292.
Recherches sur la végétation . . . Deuxième mémoire.	*C. R.*, 1854, *39*, 601–612. *A. S. N.*, 1854, *2* (Bot.), 357–370. *A. C.*, 1855, *43*, 149–223; 225.
Recherches sur la végétation. Troisième mémoire: De l'action du salpêtre sur le développement des plantes.	*C. R.*, 1855, *41*, 845–857. *J. A. P.*, 1855, *4*, 512–517. *J. A. P.*, 1856, *5*, 64–65. *A. C.*, 1856, *46*, 5–41; 225.
Remarques à l'occasion de l'extrait d'un Mémoire de M. G. Ville sur l'absorption de l'azote de l'air par les plantes, imprimé dans le Compte rendu du 10 avril 1854.	*C. R.*, 1854, *38*, 717–718.
Sur l'opportunité de faire intervenir, dans quelques circonstances, l'arsenic dans le chaulage des grains.	*A. C.*, 1856, *46*, 458–472. *B. S. A.*, 1855–56, *11*, 198–212. *J. A. P.*, 1856, *5*, 140–145.
Remarkques à l'occasion d'une communication de M. G. Ville concernant l'influence exercee sur la végétation par le phosphate de chaux des engrais.	*C. R.*, 1857, *45*, 996–1000.
Recherches sur l'influence que l'azote assimilable des engrais exerce sur la production de la matière végétale.	*C. R.*, 1857, *44*, 940–954. *C. R.*, 1857, *45*, 833–845. *J. A. P.*, 1857, *7*, 473–479. *A. S. N.*, 1857, *7* (Bot.), 5–20.
Alimentation des plantes. Influence du phosphate de chaux des engrais sur la production végétale.	*J. A. P.*, 1857, *8*, 441–449.
Recherches sur les quantités de nitrate contenues dans le sol et dans les eaux.	*C. R.*, 1857, *44*, 108–119. *A. S. N.*, 1857, *7* (Bot.), 21–34. *J. A. P.*, 1857, *7*, 105–110.
Remarques à l'occasion d'un mémoire de	*A. C.*, 1857, *50*, 479–480.

M. Mayer, inséré dans le cahier de juin 1857 des *Annales de chimie et de physique*, sur les rapports de l'acide phosphorique à l'azote dans les graines des céréales.	
Sur la quantité d'ammoniaque contenue dans la rosée artificielle.	*C. R.*, 1857, *44*, 1033–1036.
Sur l'action du salpêtre employé comme engrais.	*J. A. P.*, 1858, *2*, 446–449.
Statique des cultures industrielles de l'Alsace; ler Mémoire: le Tabac.	*C. R.*, 1858, *46*, 1007–1019. *A. C.*, 1866, *9*, 51–80. *A. C. A. M.*, 1866–67, 7.
Recherches sur la quantité d'acide nitrique contenue dans la pluie, le brouillard, la rosée.	*C. R.*, 1858, *46*, 1123–1131; 1175–1183.
Nouvelles observations sur le développement des Hélianthus soumis à l'action du salpêtre donné comme engrais.	*C. R.*, 1858, *47*, 807–813.
Du terreau et de la terre végétale.	*J. A. P.*, 1859, *1er semestre*, 493–497.
De la terre végétale considerée dans ses effets sur la végétation.	*C. R.*, 1859, *48*, 303–318; 657–667. *A. S. N.*, 1859, *12* (Bot.), 354–372. *J. A. P.*, 1859, *1er semestre*, 241–248.
De la constiution du terreau comparée à celle de la terre végétale.	*C. R.*, 1859, *48*, 931–940.
Sur la présence des nitrates dans le guano.	*C. R.*, 1860, *50*, 887–890. *J. P.*, 1860, *37*, 432–436.
Etudes sur le chaulage des terres arables.	*C. R.*, 1861, *53*, 129.
Nouvelles recherches concernant l'action exercé sur l'atmosphère par les parties verte des végétaux.	*A. C. A. M.*, 1862, *2*, 217. *C. R.*, 1861, *53*, 862.

Chemistry

Recherches sur les variations que l'eau de la mer Morte paraît subir dans sa composition.	*A. C.*, 1856, *48*, 129–170. *C. R.*, 1856, *42*, 1230–1258.
Sur l'obstacle apporté par la silice à la reconnaissance des phosphates dans les matières terreuses.	*J. A. P.*, 1859, *1er semestre*, 154–155.
Sur la présence de l'acide nitrique dans le bioxyde de manganèse.	*C. R.*, 1860, *50*, 890–892.

Various

Sur la sensation de chaleur que produit le gaz acide carboniquc sur la pcau.	*C. R.*, 1855, *40*, 1006–1009. *A. C.*, 1855, *44*, 204–209.

(e) 1860–1883

South America

Sur la composition du pulque, boisson fermentée, préparée avec la sève du maguey.	*A. C.*, 1866, *7*, 429–449. M. A. C., 1865, *1*, 209–244. *A. C. A. M.*, 1864, *5*, 489–501.
Sur la nitrière de Tacunga, Etat de l'Equateur.	*C. R.*, 1864, *59*, 218–223. *A. C.*, 1866, *7*, 358–375. *A. C. A. M.*, 1864, *5*.
Note sur la composition de la sève de l'Agave.	M. A. C., 1867, *3*, 10–11.
Sur les eaux acides qui prennent naissance dans les volcans des Cordilières.	*A. C.*, 1874, *2*, 76–130. *C. R.*, 1874, *78*, 453–461; 526–533; 593–599.
Observations sur la production du sucre des Agaves.	*C. R.*, 1875, *81*, 1070–1071.
Sur la composition du lait de l'arbre de la vache (Brosimum galactotendron).	*A. C.*, 1878, *15*, 180–184. *G. R.*, 1878, *87*, 277–281. *J. P.*, 1878, *28*, 361–364.
Détermination de la hauteur du mercure dans le baromètre sous l'équateur; amplitude des variations diurnes barométriques à diverses stations dans les Cordillères.	*C. R.*, 1879, *88*, 1158–1165. *A. C.*, 1880, *21*, 5–71.
Les sources thermales de la chaine du littoral du Vénézuéla (Amérique méridionale).	*C. R.*, 1880, *91*, 836–841. *A. C.*, 1881, *22*, 145–152.
Sur les matières sucrées contenues dans le fruit du caféier.	*A. C.*, 1879, *4*, 1–4. *J. P.*, 1880, *2*, 465–469. *C. R.*, 1880, *91*, 639–642. *A. C.*, 1881, *22*, 289–293.
Sur les outils en bronze employés par les mineurs du Perou.	*C. R.*, 1883, *96*, 545–546.
Sur la culture du cacaoyer. Recherches sur la constitution des fèves de cacao et du chocolat.	*C. R.*, 1883, *96*, 1395–1399. *J. P.*, 1883, *8*, 20–24.
Le cacao et le chocolat.	*A. C.*, 1883, *28*, 433–456.

Crop Husbandry

Observations relatives au développement des Mucédinées.	*C. R.*, 1860, *51*, 671–674. *A. C.*, 1861, *61*, 363–367. *A. C. A. M.*, 1866–1867, *7*, 5–11.
Remarques à l'occasion d'un Mémoire de M. Pasteur sur le mode de nutrition des Mucédinées.	*C. R.*, 1860, *51*, 711.
Sur la nitrification de la terre végétale	*C. R.*, 1873, *76*, 22–29. *A. C.*, 1873, *29*, 186–206. *B. S. A.*, 1872–73, *8*, 207–217.
Substances alimentaires conservées par l'action du froid.	*C. R.*, 1873, *76*, 189. *B. S. A.*, 1872–73, *8*, 266–268.
Du concours des roches volcaniques à la formation et à la fertilité de la terre végétale.	*A. C.*, 1874, *3*, 390–417.
Analyses comparées du biscuit de gluten et de quelques aliments féculents.	*A. C.*, 1875, *5*, 114–128.
Expériences pour constater la perte en sucre dans le sucrage du moût et du marc de raisin.	*A. C.*, 1876, *7*, 433–451.
Influence de la terre végétale sur la nitrification des matières organiques azotées employées comme engrais.	*A. C.*, 1876, *8*, 5–24. *C. R.*, 1876, *82*, 477–479.

Animal Husbandry

Sur la composition des fromages	*A. C.*, 1865, *3*, 398–400. *J. A. P.*, 1865, *1*, 343.
Expériences sur le barattage suivies de recherches analytiques entreprises pour constater si la nature des aliments consommés par la vache influe sur la proportion de beurre contenue dans le lait.	*A. C.*, 1866, *9*, 108–161. *A. C. A. M.*, 1866–67, 7.
Aspect du lait vu au microscope, avant et après le barattage et l'écrémage.	*A. C.*, 1872, *25*, 382–386.
Remarques sur le procédé de M. Bourrel pour empêcher la transmission de la rage.	*B. S. A.*, 1873–74, *9*, 511–512.
Recherches du fer dans le sang d'un animal invertébré.	*C. R.*, 1872, *75*, 173–174.
Observations critiques sur l'emploi de la teinture ou de la poudre de gaiac pour apprécier la pureté du kirschenwasser.	*C. R.*, 1874, *79*, 832–836. *A. C.*, 1875, *4*, 285–291.

Dosage des nitrates et de l'ammoniaque dans l'eau de la Seine, prise le 18 mars 1876 au-dessous du pont Austerlitz.	*C. R.*, 1876, *82*, 658.
Sur les quantités de nitrates et d'ammoniaque contenus dans les eaux des fleuves.	*B. S. A.*, 1876, *36*, 203–205.
Sur la décomposition du bioxyde de baryum dans le vide, à la température du rouge sombre.	*C. R.*, 1877, *84*, 521–526. *A. C.*, 1880, *19*, 464–472.
Recherches sur la présence de l'acide nitrique et de l'ammoniaque dans les eaux et la neige, recueillies dans les glaciers des Alpes par M. Civiale.	*C. R.*, 1882, *95*, 1121–1123.
Sur la composition des substances minérales combustibles.	*A. C.*, 1883, *29*, 363–392. *C. R.*, 1883, *96*, 1452–1456. *J. P.*, 1883, *8*, 249–252.

Metallurgy

Sur un procédé pour constater la présence des azotures dans l'acier, la fonte et le fer.	*C. R.*, 1861, *52*, 1008–1011.
Observations relatives au dosage de l'azote dans le fer et l'acier.	*C. R.*, 1861, *52*, 1249–1252. *C. R.*, 1861, *53*, 5–10.
Sur la présence de l'azote dans un fer météorique.	*A. C.*, 1861, *63*, 336–343. *C. R.*, 1861, *53*, 77–79.
Analyse d'une fonte chromifère. Dosage du carbone dans la fonte, le fer et l'acier.	*C. R.*, 1868, *66*, 873–877. *A. C.*, 1870, *19*, 78–118. *A. C. A. M.*, 1868, *8*.
Du dosage du graphite dans le fer carburé.	*A. C.*, 1870, *20*, 243–250.
Dosage du silicium dans la fonte, le fer et l'acier, par la voie sèche.	*A. C.*, 1871, *22*, 457–472.
Etudes sur la transformation du fer en acier par la cémentation.	*A. C.*, 1875, *5*, 145–265. *C. R.*, 1874, *78*, 1458–1464; 1513–1514.
Sur la limite de la carburation du fer.	*C. R.*, 1875, *80*, 850–858.
Sur la siliciuration du platine et de quelques autres métaux.	*C. R.*, 1876, *82*, 591–596; 1007.
Sur les cristaux d'oxyde de fer magnétique formés pendant le grillage d'un minéral spathique.	*C. R.*, 1876, *83*, 1007–1008.
Sur la production et la constitution des aciers chromés.	*C. R.*, 1878, *86*, 1303. *A. C.*, 1878, *15*, 91–126.
Sur la carburation du nickel par voie de cémentation.	*C. R.*, 1878, *86*, 509–513.

Geology

Sur la cause qui détermine la tuméfaction de l'obsidienne à une température élevée (En commun avec M. Damour).	*A. C.*, 1873, *29*, 543–564. *C. R.*, 1873, *76*, 1158–1163.
Sur l'apparition du manganèse à la surface des roches.	*A. C.*, 1882, *27*, 289–311. *C. R.*, 1882, *95*, 318–324; 368–373.

Botany

Nouvelles recherches concernant l'action exercée sur l'atmosphère par les parties vertes des végétaux.	*C. R.*, 1861, *53*, 862–884.
Sur la nature des gaz produits pendant la décomposition de l'acide carbonique par les feuilles exposées à la lumière.	*A. S. N.*, 1862, *16* (Bot.), 5–27.
Expériences entreprises pour rechercher s'il y a émission de gaz azote pendant la decomposition de l'acide carbonique par les feuilles. Rapport existant entre le volume d'acide décomposé et celui de oxygène mis en liberté.	*A. C.*, 1862, *66*, 295–429.
Lettre sur la non-émission du gaz azote dans la décomposition de l'acide carbonique par les feuilles.	*C. R.*, 1863, *57*, 412–414.
Sur l'apparition du gaz oxyde de carbone pendant l'absorption de l'oxygène par certaines substances vegetales.	*C. R.*, 1863, *57*, 885–894.
Sur la disparition des gaz combustibles mêlés à l'oxygène pendant la combustion lente du phosphore.	*C. R.*, 1864, *58*, 777–788.
De la végétation dans l'obscurité.	*C. R.*, 1864, *58*, 881–885; 917–922. *A. C.*, 1868, *13*, 219–240. *A. S. N.*, 1864, *1* (Bot.), 314–324.
Etudes sur les fonctions des feuilles.	*C. R.*, 1865, *60*, 872–880. *C. R.*, 1865, *61*, 493–506; 605–613; 657–665. *A. C.*, 1868, *13*, 282–416. *A. S. N.*, 1869, *10* (Bot.), 331–343. M. S. A., 1865, *111*, 344–392.

Sur les fonctions des feuilles.	*C. R.*, 1866, *63*, 706–714; 748–756. *C. R.*, 1869, *68*, 410–420. *A. C.*, 1869, *18*, 321–352.
Sur l'action délétère que la vapeur émanant du mercure exerce sur les plantes.	*C. R.*, 1867, *64*, 924–929; 983–989.
Sur une matière sucrée apparue sur les feuilles d'un tilleul.	*C. R.*, 1872, *74*, 87–90. *A. C.*, 1872, *25*, 5–21. *B. S. A.*, 1871–72, *7*, 172–176.
Réponse à une Communication de M. Harting sur le même sujet.	*C. R.*, 1872, *74*, 473.
Sur la sorbite, matière sucrée analogue à la mannite, trouvée dans le jus des baies du sorbier des oiseleurs.	*C. R.*, 1872, *74*, 939.
Observations sur la production du sucre des Agaves.	*C. R.*, 1875, *81*, 1070–1071.
Végétation du mais commencée dans une atmosphère exempte d'acide carbonique.	*C. R.*, 1876, *82*, 788–791. *A. C.*, 1876, *8*, 433–443.
Sur la végétation des plantes dépourvues de chlorophylle.	*C. R.*, 1876, *82*, 939–943.
Etudes sur les fonctions physiques des feuilles: transpiration, absorption de la vapeur aqueuse, de l'eau, des matières salines.	*A. C.*, 1878, *13*, 289–394.
Sur la dissociation de l'acide des nitrates pendant la végétation accomplie dans l'obscurité.	*A. C.*, 1882, *22*, 433–450. *B. S. A.*, 1881, *41*, 336–341.

Various

Sur la sensation de chaleur que le gas acide carbonique fait naître dans son contact avec la peau.	*A. C.*, 1866, *8*, 381–388. *A. C. A. M.*, 1866–67, *7*.
Sur la congélation de l'eau.	*C. R.*, 1871, *73*, 77–79. *A. C.*, 1872, *26*, 544–547.
Observations sur la température de la grêle prise immédiatement après sa chute.	*C. R.*, 1880, *89*, 202. *A. C.*, 1884, *3*, 425–429.

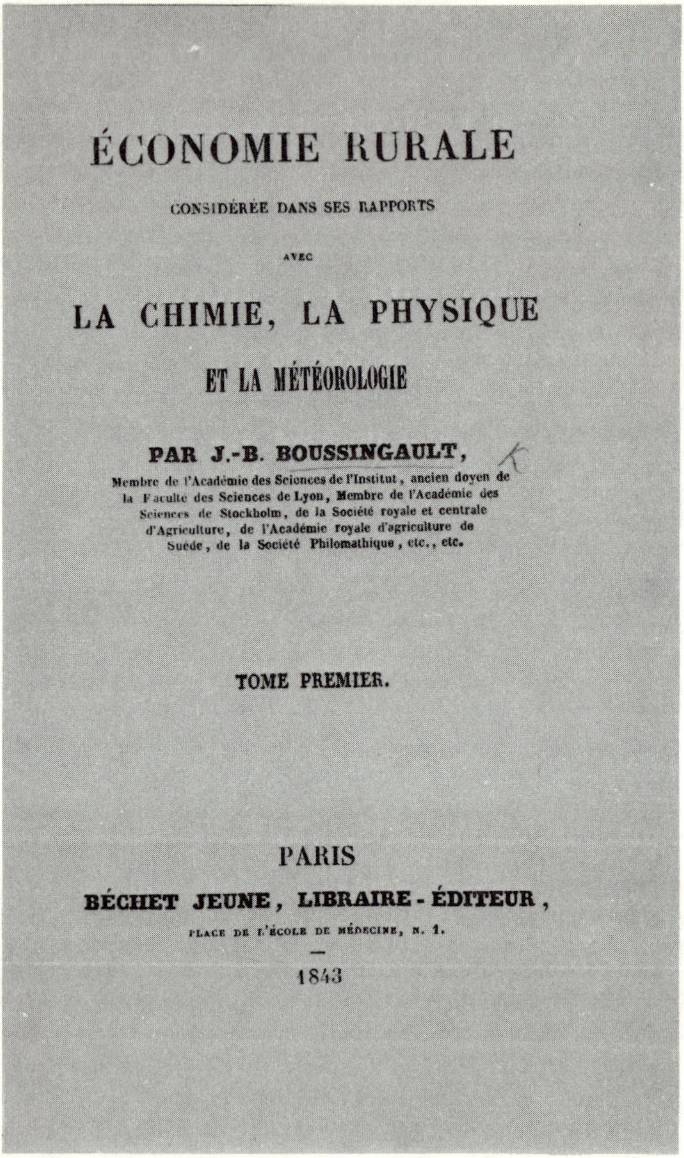

ÉCONOMIE RURALE

CONSIDÉRÉE DANS SES RAPPORTS

AVEC

LA CHIMIE, LA PHYSIQUE

ET LA MÉTÉOROLOGIE

PAR J.-B. BOUSSINGAULT,

Membre de l'Académie des Sciences de l'Institut, ancien doyen de la Faculté des Sciences de Lyon, Membre de l'Académie des Sciences de Stockholm, de la Société royale et centrale d'Agriculture, de l'Académie royale d'agriculture de Suède, de la Société Philomathique, etc., etc.

TOME PREMIER.

PARIS

BÉCHET JEUNE, LIBRAIRE - ÉDITEUR,

PLACE DE L'ÉCOLE DE MÉDECINE, N. 1.

—

1843

Fig. 31. Title page of Boussingault's *Economie rurale.* (By kind permission of the British Library.)

WORKS OF BOUSSINGAULT PUBLISHED IN BOOK FORM

Economie rurale considérée dans ses rapports avec la chimie, la physique et la météorologie, Paris, 1843–1844, 2 vols, 2nd ed., 1851.

Die Landwirthschaft in ihren Beziehungen zur Chemie, Physik und Meteorologie, trans. Dr. A. Graeger, Halle, 1844–1845, 2 vols.

Rural Economy, trans. G. Law, London, 1845.

Essai de statique chimique des êtres organisés (with Dumas), Paris, 3rd ed., 1844.

The Chemical and Physiological Balance of Organic Nature (with Dumas), trans. G. Law, London, 1844.

Recherches sur le moyen d'extraire, pour les besoins des arts, le gaz oxygène contenu dans l'air atmospherique, Paris, 1852 (reprinted from *Annls chim. phys.*, 1852, *35*, 5–54).

Mémoires de chimie agricole et de physiologie, Paris, 1854.

La fosse à fumier, Paris, 1858.

Agronomie, chimie agricole et physiologie, Paris, 2nd ed., 1860–1874, 5 vols.; 3rd ed. 1886–1891, 8 vols.

Observations sur l'avis donné par M. Salvelat, en réponse aux questions posées par la Compagnie parisienne par MM. Boussingault . . . Gallien . . . Wurtz, Paris, 1865.

Dosage du carbone dans la fonte, le fer et l'acier, Paris, 1870 (reprinted from *Annls chim. phys.*, 1870, *19*, 78–118).

Etudes sur la transformation du fer en acier par la cémentation, Paris, 1875 (reprinted from *Annls Chim. Phys.*, 1875, *5*, 145–265).

Analyses comparées du biscuit de gluten et de quelques aliments féculents, Paris, 1875 (reprinted from *Annls chim. phys.*, 1875, *5*, 114–128). This work is mentioned by Franqueville, Comte de, *Le premier siècle de l'Institut*, Paris, 1895, *1*, 257.

Sur les matières sucrées contenues dans le fruit du Caféier, Paris, 1881 (reprinted from *Annales de l'Institut national agronomique*, No. 4, 1878–79).

Les secousses souterraines dans les Andes, Paris, 1887 (reprinted from *Agronomie, chimie agricole et physiologie*, Vol. 8).

Mémoires de J. B. Boussingault, Paris, 1892–1900, 5 vols.

L'air, l'acide carbonique et l'eau, Paris, 1913, one of *Les classiques de la science*. (It contains 'Recherches sur la véritable constitution de l'air atmosphérique', *Annls chim. phys.*, 1841, *3*, 257–305.)

DOCUMENTARY MATERIAL

1. Archives de l'Académie des Sciences

Boussingault Dossier

To Meynard de la Groye, Nov. 1819.
To J. B. Dumas, 38 letters, 1839–1881.
To Hervé Mangon, 3rd Dec., 1886.
A précis of the correspondence (now missing) between Boussingault and the Rector of Lyon, and between the Rector and Guizot. Précis made by Prof. A. Offert at Lyon in 1925.
From Mme Alice Crozet-Boussingault to M. Pierre Gauja, July 1926.
Edmond Boussingault to Alfred Lacroix, 14th April, 1941, stating his intention of publishing the letters of Humboldt to Boussingault' 1822–1859 (see Humboldt dossier). Edmond Boussingault to Alred Lacroix, 12th May, 1941, mentioning that he has numerous letters from Boussingault to Fourneyron, Liebig and Regnault (these have not been traced).
Ordonnance du Roi, Louis Phillipe, of 31st January, 1839, on the election of Boussingault to the Académie Royale des Sciences.
A small album containing about six photographs depicting a visit to Bechelbronn and Liebfrauenberg on July 7th, 1934, under the auspices of the Association Internationale de la Science du Sol.
'Notice sur la Mémoire et observations relatifs à l'Agriculture, la Météorologie et la Physique du Globe, etc.' On the flyleaf is written, "Boussingault – études 1827–1837". These are four papers written on his South American experiences and ends with a note: "In submitting this account of his works to the Académie, M. Boussingault feels that he can add that since his return to Europe he has been almost exclusively occupied with practical agriculture on his family's farm".
3 letters from Boussingault to Peligot, 1853–54.

J. B. Dumas Dossier

Carton 16, file "Enseignement, Conservatoire des Arts et Métiers". Letter from A. Delambre, temporary head of the Division of Commerce to Minister of Agriculture and Commerce, about Febrary 1850.
Carton 18, advertisements of courses at the Sorbonne, 1836–1845 in which there is no mention of Boussingault.
"Commission des Monnaies 1867–70".

Ville Dossier

Portion of notes on the nitrogen assimilation experiments conducted before a committee of the Académie des Sciences at the Muséum d'histoire naturelle; also a page of results.

Roulin Dossier

Photograph of a self-portrait, in oils of Roulin, which belonged to M. Georges Bertrand, Consul Technique of the Institut de France.

von Humboldt Dossier

33 letters from von Humboldt to Boussingault, 1822–1858. These are typed copies and are referred to in the Boussingault dossier in two letters from Edmond Boussingault to Alfred Lacroix.

2. Bibliothèque de l'Instiut de France

MS 1–14 (2420–2433) of which letters B359–384 are from Boussingault to Mabru (Director of Mines, Pechelbronn), Daubrée (professor of geology at Strasbourg) and Voltz (Chief Mining Engineer, Strasbourg) from 1823 to 1876. B 385 is the Boussingault funeral notice.

MS 15 (2043) letter 48 from Boussingault to François Arago, 13th July, 1847.

MS 3817 (6) 'Deuxième mémoire sur l'agriculture des tropiques – la Cacao'. This was published in *Comptes Rendus*, 1836, *3*, 510–513, but whole paragraphs have been omitted.

3. Bibliothèque centrale du Muséum d'histoire naturelle

MS 2642, letters 65–71, correspondence between Boussingault and Ville, 1848–49.

4. Archives du Conservatoire des Arts et Métiers

Procès-Verbaux du Conseil de Perfectionnement 1845–1887.

Correspondence, May 1876–May 1880 over transfer of Institut agronomique from Versailles to the Conservatoire, from Morin to Minister of Agriculture and Commerce, 1876 – 5 letters.

Correspondence between Joseph Boussingault and Director of the Conservatoire; letter from Minister of Public Works to Minister of Commerce and Industry – three letters concerning the statue of Boussingault, 1892 and 1904.

5. Archives du Conseil d'Etat

Folders 17011, 15429, 12081, 10, 604–605, 9669, 8069, 8427, of the period 1850–51. This is all that survives from the fire of 1871.

6. Archives de la Garantie du Ministère des Finances, rue de Perrée, Paris

Handwritten exercise book labelled 1899–1904 containing letters on assays signed by Joseph Boussingault.

7. Château Holtzer, Unieux (Haut-Loire)

Letters to Mme Holtzer from MM. Brustlein and Kopp. A notebook used mainly by Joseph Boussingault.

8. Bibliothèque municipale de Lyon

MS 363089 Académie de Lyon. Installation de la faculté des Sciences de Lyon, 1835.

Copies of *Le Courier de Lyon, Le Réparateur* and *l'Athenée* for the period 1834–1836.

9. Lampertsloch (Alsace), Registers of Births and Marriages

Marriage record of Boussingault and Mlle LeBel. Birth records of Berthe and Joseph Boussingault.

10. Archives de Bas-Rhin, Strasbourg

MS 7M 489, Etat nominatif des habitants – commune de Lampertsloch, 1836, 1841, 1846, 1851, 1856, 1861, 1866.

MS 7M 389, Etat nominatif des habitants – commune de Goersdorf, dates as for Lampertsloch.

MS 2M 73, Réunion Générale du Comité Central de Strasbourg et des délégues des comités cantonaux du département du Bas-Rhin. Procés-verbal de la séance du 19 mars 1848.

Election results for the various cantons of the département of Bas-Rhin, 1848.

11. Bibliothèque Nationale et Universitaire, Strasbourg

MS R 5173–5174 two notebooks of Boussingault, 'H', 1859–60; 'Z', 1864.
BRANTHÔME Abbé, *Chimie théorique et pratique*, 1826–27, notes of lectures by Charles Schmidt bound into one volume, MS 3177 (557) 19.

12. Archives de l'Académie de Paris

Registre des Procès-Verbaux des Actes et délibérations de la Faculté des Sciences de l'Académie de Paris, 26th October, 1835.
Letter to dean of the faculty of Sciences from the Minister of Public Instruction, 21st March, 1837.

13. Historisches Archiv., der Magistrat der Stadt Wetzlar

Marriage record of Boussingault's parents, 8th September, 1796.

LIST OF WORKS CONSULTED

Unless otherwise stated, the works mentioned below have been published in Paris or London.

Alexander, W. and Street, A.: *Metals in the Service of Man* (Harmondsworth, 1944).

Anschutz, R.: *August Kekulé* (Berlin, 1929), Vol. 1, Appendix 9.

Artz, F. B.: *The Development of Technical Education in France 1500–1850* (Cambridge, Mass., 1966).

Audoc, L.: *Le Conseil d'Etat* (1876).

Aulie, R. P.: *Boussingault and the Nitrogen Cycle*, Ph.D. thesis, Yale (1968).

Aulie, R. P.: 'Boussingault and the nitrogen cycle', *Proceedings of the American Philosophical Society*, 1970, *114*, No. 6, 435–479.

Aulie, R. P.: 'Origin of the idea of the nitrogen cycle', *The American Biology Teacher*, 1971, *33*, No. 11.

Aulie, R. P.: 'The mineral theory', *Agricultural History*, 1974, *48*, No. 3, 369–382 (Univ. California Press).

Baker, J. N. L.: *A History of Geographical Discovery and Exploration*, 1931, reprinted 1948, 404.

Barclay, D.: 'Experiments with nitrate of soda', *Journal of the Royal Agricultural Society*, 1840, *1*, 423–429.

Bard, J.: *Le palais St. Pierre ou des Arts* (Lyon, 1842).

Berman, M.: *Social Change and Scientific Organization: The Royal Institution 1799–1844* (1978).

Bourde, A. J.: *The Influence of England on the French Agronomes 1750–1789* (Cambridge, 1953), Chapter 6.

Boussingault, J. B.: *Economie rurale*, 2nd ed. 1851, 2 vols.; trnas. G. Law, *Rural Economy*, 1845, 1 vol.

Boussingault, J. B.: *La fosse à fumier*, 1858.

Boussingault, J. B.: *Agronomie, chimie agricole et de physiologie*, 3rd ed., 886–1891, 8 vols.

Boussingault, J. B.: *Funérailles de M. Dutrochet*, Académie des Sciences, 1847.

Boussingault, J. B.: *Mémoires de J. B. Boussingault*, 1892–1903, 5 vols.

Browne, C.: *Chronica Botanica* (Waltham, 1944), Vol. 8.

Buckett, M.: *Introduction to Livestock Husbandry* (Oxford, 1957).

Cannizzaro, S.: *Il nuovo cimenta* (Genoa, 1858) trans. as *Sketch of a Course of Chemical Philosophy*, Alembic Club reprint No. 18 (Edinburgh, 1910).

Carr, J. C. and Taplin, W.: *History of the British Steel Industry* (Oxford, 1962), 132.

Chambrier, P. de: *Historique de Pechelbronn 1498–1918* (Neuchâtel, 1919).

Chambrier, P. de: *Exploitation du pétrole par puits et galeries*, 1921.
Chambrier, P. de: *Les mines de pétrole de Pechelbronn* (Strasbourg, 1920).
Clow, A. and Clow, N. L.: *The Chemical Revolution*, 1952, 502.
Cochranc, C. S.: *Journal of a Residence and Travels in Colombia During the Years 1823–1824*, 1825, 2 vols.
Combes, M.: *Roulin et ses amis 1796–1874*, 1928.
Cordier, J.: *Mémoire sur l'agriculture de la Flandre française*, 1823.
Cowgill, G. R.: 'Jean Baptiste Boussingault – a biographical sketch', *Journal of Nutrition*, Philadelphia, 1964, *84*, 1–9.
Dana, S. L.: *A Muck Manual for Farmers* (Lowell, 1842).
Deherain, P. P.: 'L'oeuvre agricole de M. Boussingault', *Revue scient.* 1887, *2*, 33–46.
Delépine, M.: *Vie et oeuvres de Joseph Achille LeBel*, 1949.
Demolon, A.: *L'évolution scientifique et l'agriculture français*, 1946.
Denison, J. E.: 'Report on the agricultural department of the Paris Exhibition', *Journal of the Royal Agricultural Society*, 1856, *17*, 58.
Descola, J.: *Les messagers de l'independence*, 1973.
Dumas, J. B. and Boussingault, J. B.: *Essai de statique chimique des êtres organisés*, 3rd ed. 1844, trans. G. Law, *The Chemical and Physiological Balance of Organic Nature*, 1844.
Dutacq, F.: 'Les origines de l'université de Lyon', *Annales de l'université*, Lyon, 1939–1940.
Ehrhard, A.: *L'université de Lyon* (Lyon, 1919).
Fohlen, C.: *Histoire de Besançon* (1964–65), Vol. 1.
Fox, R. and G. Weisz (eds.): *The Organization of Science and Technology in France 1808–1914*, 1980.
Frazier, W. C.: *Food Microbiology* (New York, 1958), 302–303.
Freedeman, C. E.: *The Conseil d'Etat in Modern France* (New York, 1961).
Fresenius, C. R.: *Quantitative Chemical Analysis*, trans. C. E. Groves from 6th ed., 1900, Vol. 2.
Grasse, P. (ed.): *Traité de zoologie*, 1955, *17*, 1106.
Green, J. R.: *A History of Botany 1860–1900* (Oxford, 1909).
Grouvelle, P.: *Chauffage et ventilation de la Nouvelle Force à Paris*, 1845.
Hadfield, R. A.: *Metallurgy and Its Influence on Modern Progress*, 1925.
Hall, A. D.: *The Book of the Rothamsted Experiments*, 1905.
Hamilton, J. P., *Travels Through the Interior Provinces of Colombia*, 1827, 2 vols.
Hamy, C. T.: *Correspondence d'Alexandre de Humboldt avec François Arago 1809–1853*, 1907.
Hervier, M.: *Les palais des arts* (Lyon, 1922).
Hillairet, J.: *Evocation de vieux Paris*, 1952, Vol. 1.
Howard, M.: *The Franco-Prussian War*, 1961.
Humboldt, A. V.: *Examen critique de l'histoire de la géographie du nouveau continent et des progrès de l'astronomie nautique aux quinzième et seizième siècles*, 1836–1839, xx–xxi.
Humphries, R. A.: 'The emancipation of Latin America', *The New Cambridge Modern History* (Cambridge, 1965), Vol. 9.

Johnson, D.: *Guizot*, 1963.

Johnson, J. J.: *Simon Bolivar and Spanish American Independence 1783–1830* (Princeton, 1968).

Kellner, C.: *Alexander von Humboldt*, 1963.

Kellner, O.: *The Scientific Feeding of Animals*, trans E. Goodwin, 1908, 2nd ed. 1926, 183.

Kisker, G. W.: *The Disorganised Personality* (New York, 1964).

Kolthoff, I. M. and Elving, P. G.: *A Treatise on Analytical Chemistry* (New York, 1961). (Vols 2, 5 and 7 of Part 2 have been consulted for a comparison of Boussingault's methods with those of today.)

Kuhlmann, C. F.: 'Expériences sur la fertilisation des terres par les sels ammoniacaux, les nitrates et d'autres composés azotes', *Comptes rendus*, 1843, *17*, 1118–1128.

Kuhlmann, C. F.: Expériences concernant la théorie des engrais', *Annls chim. phys.*, 1846, *18*, 138–157.

Kuhlmann, C. F.: 'Relation entre la nitrification et la fertilisation des terres', *Annls chim. phys.*, 1847, *20*, 224–225.

Lacroix, A.: *Notice historique sur Jean Baptiste Boussingault*, 1926.

Lacroix, A.: *Minéralogie de la France*, 1962.

Lacroix, A.: *Figures de savants*, 1937, 2, 115–177. (This is a reprinting of the author's *Notice historique . . . .*)

Ladenburg, A.: *Lectures on the History and Development of Chemistry Since the Time of Lavoisier*, trans. L. Dobbin, revised ed. (Edinburgh, 1905).

Lavisse, E.: *Histoire de France contemporaine*, 1920, Vol. 2.

Lavoisier, A. L.: *Traité élémentaire de chimie*, 1789.

Lavoisier, A. L.: *Instruction sur la fabrication du salpêtre*.

Lavollay, J.: 'Aspects et étapes de la recherche agronomique en France', No. 28, *Comptes rendus des séances de l'Académie d'Agriculture*, 1961, *47*, 233–239.

Lawes, J. B. and Gilbert, J. H.: 'Agricultural chemistry', *Jl R. agric. Soc.*, 1847, *8*, 226.

Lawes, J. B. and Gilbert, J. H.: 'Agricultural chemistry', *J. R. agric. Soc.*, 1851, *12*, 1.

Lawes, J. B. and Gilbert, J. H.: *Report of the British Association for the Advancement of Science*, 1852, 323–353.

LeClerc-Thouin, O.: 'Notions théorique et pratique sur les assolements', *La maison rustique du 19ième siècle*, 1835, 257.

LeFebre, G.: *The French Revolution*, 1967, Vol. 2.

Lehmann, C. G.: *Physiological Chemistry*, trans. G. E. Day (Philadelphia, 2nd ed., 1855), 2 vols.

Lenglen, M.: *J. B. Boussingault à la société centrale d'agriculture* (Beauvais, 1937).

Lenglen, M.: *Boussingault à Bechelbronn et au Liebfrauenberg* (Alencon, 1940).

Leonard, C. M.: *Lyon Transformed* (Berkeley, 1961).

Leouzen, L.: *Agronomes et éleveurs*, 1905, 317–331.

LeRoux, D.: 'J. B. Boussingault créateur de la chimie agricole. Sa vie et son oeuvre au Conservatoire National des Arts et Métiers', *Bulletin de L'association française pour l'étude du sol*, 1936, *2*.

Liebig, J. v.: *Organic Chemistry in Its Application to Agriculture and Physiology*, trans. L. Playfair, 1840, also 3rd ed., 1843 and 5th ed., 1846.

Liebig, J. v.: *Familiar Letters on Chemistry and Its Relation to Commerce, Physiology and Agriculture*, ed. J. Gardner (New York, 1843).

Liebig, J. v.: *The Natural Laws of Husbandry*, trans. J. Blyth, 1963.

Lucas-Dubreton, J.: *The Restoration of the July Monarchy*, trans. E. F. Buckley, 1929.

Maclaran, M. J.: *Gold, Its Geological Occurrence and Geographical Distribution*, 1908.

McCollum, E. V.: *A History of Nutrition* (Boston, 1957).

McCosh, F. W. J.: 'Boussingault versus Ville: The Social, Political and Scientific Aspects of their Disputes', *Annals of Science*, 1975, *32*, 475–490.

McCosh, F. W. J.: 'Jean Baptiste Boussingault and Platinum', *Platinum Metals Review*, 1977, *21*, No. 3, 97–100.

McCosh, F. W. J.: 'Malaria and Goitre in South America One Hundred and Fifty Years Ago', *Cent. Afr. Jl Medicine*, 1977, *23*, No. 11, 254–258.

McDonald, D.: *A History of Platinum*, 1960.

McDonald, D. and L. B. Hunt: *A History of Platinum and Its Allied Metals*, 1982.

McIntyre, A. R.: *Curare: Its History, Nature and Chemical Use* (Chicago, 1947).

McKee, H. S.: *Nitrogen Metabolism in Plants* (Oxford, 1962).

Madariaga, S. de: *Bolivar*, 1952.

Magendie, F.: *An Elementary Treatise on Human Physiology*, 5th ed. (New York, 1844(, trans. J. Rever from *Précis élémentaire de physiologie*, 1811–1817.

Magendie, F.: 'Mémoire sur les propriétés nutritives des substances qui ne contiennent pas d'azote', *Annls chim. phys.*, 1816, *3*, 68–77.

Malpighi, M.: *Anatome plantarum* (London, 1687).

Mariotte, E.: *Oeuvres de M. Mariotte* (Leiden, 1717).

Mellor, J. W.: *A Comprehensive Treatise on Inorganic and Theoretical Chemistry*, 1957, imp. Vols. 8, 12 and 13.

Meyer, E. V.: *A History of Chemistry*, trans. G. McGowan, 1891.

Meyer-Abich, A.: *Alexander von Humboldt 1769–1969* (Bad-Godesberg, 1967), in English.

Monceau, Duhamel: *Eléments d'agriculture* 1762, 2 vols.

Montarzueil, G.: *Les eaux et les fontaines de Paris*, 1928.

Morrell, J. B.: 'The chemist breeders; the research schools of Liebig and Thomas Thomson', *Ambix*, 1972, 19, No. 1.

Moulton, F. R. (ed.): *Liebig and After Liebig* (Washington, 1942).

Mulder, G. J.: *Chemistry of Vegetable and Animal Physiology*, trans. P. F. H. Fromberg, 1849.

Ostwald, W.: *Les grands hommes*, 1912, trans. M. Dufour from *Grosse Männer* (Leipzig, 1909).

Palladin, V. I.: *Plant physiology*, trans. B. E. Livingston (Philadelphia, 1923).

Parias, L. H. (ed.): *Histoire universelle des explorations*, 1956, 1.

Parra, H.: 'Simple goiter in Colombia', *American Journal of Public Health*, 1948, *38*, 820–821.

Partington, J. R.: *A History of Chemistry*, Vol. 4, 1964.

Pasteur, L.: 'Mémoire sur la fermentation appelée lactique', *Comptes rendus*, 1857, *45*, 913–916.

Pasteur, L.: 'Nouveaux faits pour servir à l'histoire de la levûre lactique', *Comptes rendus*, 1859, *48*, 337–338.

Pasteur, L.: *Bulletin de la société nationale d'agriculture de France*, 1880, *40*, 815–819 (in which Pastuer pays a tribute to the analytical skill of Joseph Boussingault).

Pendle, G.: *A History of Latin America*, 1963.

Pfleger, A.: 'J. B. Boussingault und schloss Liebfrauenberg', *Elsassland* (Strasbourg, 1920), 390–393.

Phillips, J. A. and Louis, R.: *A Treatise on Ore Deposits*, 1884.

Portères, R.: 'Georges Ville (1824–1897)', *Bulletin du Muséum d'histoire naturelle*, 1957, *29*, No. 3, 295–296.

Pouzet, Ph.: *Le Passé du lycée Ampère* (Lyon, 1937).

Priestley, J.: *Experiments and Observations on Different Kinds of Air*, 1776, *1*.

Prjanischnikow, D. N.: 'Cent ans de chimie agricole', *Bulletin de l'association feançaise pour l'étude du sol*, 1936, *2*, 167–173, from *Izvestia* of 30th May, 1936.

Prothero, R. E., 1st baron Ernle: *English Farming, Past and Present*, 1912, 6th ed. 1961, 367.

Prout, W.: *Chemistry, Meteorology, etc.*, Bridgewater treatises, 1834.

Prout, W.: *Philosophic Magazine*, 1827, *2*, 144; 1828, *3*, 33 and 107, cited from E. V. McCollum, *A History of Nutrition* (Boston, 1957), 88 and 98.

Rabinowitch, E. I.: *Photosynthesis and Related Processes* (New York, 1945), *1*.

Rabinowitch, E. I. and Govindjee: *Photosynthesis* (New York, 1969).

Riehl, H.: *Tropical Meteorology* (New York, 1954), 117.

Robert, Ad. *et al.*: *Dictionnaire des Parlementaires français*, 1891, *1*.

Roberts-Austin, W. C.: *An Introduction to the Study of Metallurgy*, 1890.

Roscoe, H. F. and Schorlemmer, C.: *A Treatise on Chemistry*, 5th ed. 1913, *2*, 450–452.

Roulin, F.: *Histoire naturelle et souvenirs de voyage*, 1865, 78.

Roux, E.: *Souvenirs de Georges Ville*, Société française d'histoire d'agriculture, presented at a meeting on 26th June, 1944; A reneoed copy at Bibliothèque centrale, Muséum d'histoire naturelle.

Russell, E. J.: *Soil Conditions and Plant Growth*.

Russell, E. J.: *A History of Agricultural Science in Great Britain 1620–1954*, 1966.

Sachs, J. V.: *History of Botany (1530–1860)*, trans. H. Garnsey (Oxford, 1890), 2nd imp., 1906.

Sachs, J. V.: *Textbook of Botany, Morphological and Physiological* (Oxford,

1875), trans. A. Bennet and W. T. Dyer from *Lehrbuch der Botanik* (Leipzig, 1868).

Salet, G.: 'J. B. Boussingault', *Revue scient.*, 1888, *41*, 373.

Saussure, N. T. de: *Recherches chimiques sur la végétation*, 1804.

Schnabel, C.: *Handbook of Metallurgy*, trans. H. Louis, 1898.

Singer, C. and Underwood, E. Ashworth: *A Short History of Medicine* (Oxford, 1928).

Singer, C. *et al.*: *A History of Technology* (Oxford, 1957), Vol. 3.

Société d'histoire moderne: *Les ministères français (1788–1909)*, Série des Instruments de Travail 1910, Vol. 2.

Stiles, W.: *An Introduction to the Principles of Plant Physiology*, 1950.

Stodart, J. and Faraday, M.: 'On the Alloys of Steel', *Phil. Trans. R. Soc.*, 1822, *112*, 253.

Szabadváry, F.: *History of Analytical Chemistry*, trans. G. Svehla (Oxford, 1966).

Thaer, A. D.: *The Principles of Agriculture*, trans. W. Shaw and C. Johnson, 1844, 2 vols, from *Grundsatze der rationellen Landwirschaft* (Berlin, 1809–1812), 4 vols.

Thibaud, M.: 'Des procédés suivis à Chessy pour la préparation mécanique des minerais', *Annales des Mines*, 1827, *3*, 195–207.

Thomson, T.: *Chemistry of Animal Bodies* (Edinburgh, 1843).

Thorne, L. T.: 'A review of the methods of producing oxygen commerically', *Journal of the Society of Chemical Industry*, 1890, *9*, 246–253.

Tisserand, E.: *Notice sur les titres et travaux de M. Eugène Tisserand*, 1883.

Tisserand, E. and LeFébure, L.: *Etude sur l'économie rurale de l'Alsace*, 1869.

Trend, J. B.: *Bolivar and the Independence of Spanish America*, 1946.

Tresse, R.: 'J. A. Chaptal et l'enseignement technique de 1800 à 1819', *Revue histoire des sciences*, 1957, *10*, 167–174.

Troost, L. J.: *Funérailles de M. Boussingault*, Académie des Sciences, 14th May, 1887.

Ville, G.: *Recherches expérimentales sur la végétation*, 1853 and 1857.

Ville, G.: 'Note sur l'assimilation de l'azote de l'air par les plantes et sur l'influence qu'exerce l'ammoniaque dans la végétation', *Comptes rendus*, 1850, *31*, 578–580.

Ville, G.: 'Recherches expérimentales sur la vegetation', *Comptes rendus*, 1852, *35*, 464–468.

Ville, G.: 'Absorption de l'azote de l'air par les plantes', *Comptes rendus*, 1854, *38*, 705–709, 723–727.

[Ville, G.:] 'Le professeur Georges Ville', *La science illustrée* No. 193, 8th Aug., 1891 (Unsigned and extremely eulogistic).

Ville, G.: *Les engrais chimiques*, 1868, trans. W. Crookes as *On Artificial Manures*, 1879, 2nd ed., *Artificial Manures*, 1909.

Voelcker, A.: *Journal of the Royal Agricultural Society*, 1856, *17*, 191 (in which he demonstrates that loss of nitrogen from manure on standing is negligible).

Wells, A. D. (ed.): *Annual of Scientific Discovery for 1856* (Boston, 1856), 272.

Wackermann, C.: *Le champ pétrolifère de Pechelbronn*, 1951, Travaux du laboratoire de Géographie de l'université de Strasbourg.

White, J. H.: *The History of the Phlogiston Theory*, 1932.

Whymper, E.: *Travels Amongst the Great Andes of the Equator*, 1892.

Wolf, A.: *A History of Science, Technology and Philosophy in the Eighteenth Century*, 1938, 636–637.

Yvart, J. A. V.: *Nouveau cours complet d'agriculture théorique et pratique*, 1821.

Yvart, J. A. V.: *Notice historique sur l'origine et les progrés des assolements raisonnées, 1821.*

Almanach de l'université Royale de France, 1836–1845.

Archives de la Commission Scientifique du Mexique, vols. 1–3, 1865–1867.

Ecole Nationale Supérieure des Mines de Saint-Etienne, Annuaire (Saint-Etienne, 1965).

Inauguration du monument de Boussingault au Conservatoire des Arts et Métiers à Paris le Dimanche 7 juillet 1895.

Le centenaire de l'Ecole Nationale des Mines de Saint-Etienne (Saint-Etienne, 1921).

Les chimistes français du 19e siècle, Musée Centennal de la clásse 87 – arts chimiques et pharmacie, à l'exposition universelle internationale, 1900.

Les établissements Jacob Holtzer, c. 1930, pub. Devambez (no author, date of publication or place); a copy seen at Château Holtzer, Unieux.

Livret de l'étudiant (Lyon, 1967).

Rapport de l'administration de la Commission Impériale sur la section française de l'Exposition Universelle de 1862, 1864.

Report of the Commission for the Exhibition of 1861, (H. M. S. O., 1863).

Rapports Générale sur les travaux du Conseil d'Hygiène publique et de Salubrité du départément de la Seine, 1844–1887.

Revue du Lyonnais, 1838, *7*, 497–498.

Serials

Annales de chimie et de physique, 1821–1884.

l'Athenée, Lyon, 1835.

Comptes rendus hebdomadaire des séances de l'Académie des Sciences, 1835–1883.

Le Courrier de Lyon, 1834–1836.

Journal of the Royal Agricultural Society, 1840–1856.

Le Moniteur universelle, 1848–1849.

Procès-Verbaux du Conseil de Perfectionnement du Conservatoire des Arts et Métiers, 1845–1887 (at the Conservatoire archives).

Le Réparateur, Lyon, 1834–1836.

INDEX OF NAMES

INDEX OF SUBJECTS